SpringerBriefs in History of Science and Technology

More information about this series at http://www.springer.com/series/10085

Sreerup Raychaudhuri

The Roots and Development of Particle Physics in India

 Springer

Sreerup Raychaudhuri
Department of Theoretical Physics
Tata Institute of Fundamental Research
Mumbai, India

ISSN 2211-4564   ISSN 2211-4572   (electronic)
SpringerBriefs in History of Science and Technology
ISBN 978-3-030-80305-6   ISBN 978-3-030-80306-3   (eBook)
https://doi.org/10.1007/978-3-030-80306-3

This Springer imprint is published by the registered company Springer Nature Switzerland AG
The registered company address is: Gewerbestrasse 11, 6330 Cham, Switzerland

Preface

This preface is something of an apology, for this book is not really a work of true scholarship. True, the undersigned is a practising scientist, even a particle physicist, and so the statements about science are, on the whole, accurate. However, science merely forms the backdrop for the story, where different times, different places and different people play across the broad canvas. The material for this epic story has been collected almost wholly from the Internet. The author has not visited libraries or archives, has not pored over musty manuscripts and letters, nor even interviewed hoary elders with sharp memories, as the true historian of science is supposed to do. Living immersed in an atmosphere of science and scientists has given some perspective, no doubt, which has certainly helped to shape this volume, but otherwise, the research behind this work is rather superficial. Likewise, the opinions expressed by the author are essentially his own.

Readers are therefore warned that those who are looking for an in-depth analysis of some particular issue in the history of the development of science in India will be sorely disappointed if they look for it in this volume. On the other hand, it may be hoped that those who want a quick introduction to the broad picture of how modern science came to India and thrived, and took root in one particular area, among others, viz. particle physics and cosmic rays, will find this work more to their satisfaction. Since science and scientists function within a society, and are much more human than their public image supposes, a fair amount of social and historical background has also been introduced wherever necessary. Sometimes interesting titbits of information have been inserted, often through footnotes. Likewise, Indian science being a part of international science, occasionally scientific developments in the Western world have been described at length to put the Indian work in its proper context. It is to be hoped that readers who have no prior introduction to the subject will find this work moderately informative and interesting. Therefore, this volume is split into two parts. Those who just want to know the story of how modern science grew in India can concentrate on the first part. Those who are more interested in cosmic rays and particle physics, and the pioneers of these subjects in India, can focus on the second part. In either of these, the stress has been on telling a story, with the science explained when it is crucial. Some overall assessments were necessary, but on the whole, the author has tried not to be too judgemental in his approach.

A note about spellings. In the nineteenth century, Indian names were written in English in a phonetic way, for the ease of the English rulers. Many of these quaint spellings have been retained in this work, mostly for the old-world charm they give to these stories of old times. Many place names have changed in modern India—these are generally mentioned in brackets and the colonial names retained. Indian words have mostly been italicised, except really common ones like, e.g. pundit.

This work owes its initiation to my good friend, B. Ananthanarayan of the Indian Institute of Science in Bengaluru, and its completion to my wife, Oindrila, for allowing me to stay at my desk for totally absurd lengths of time. I am also grateful to the publishers for standing by all my idiosyncrasies, including modifying the theme of the work midway.

Mumbai, India   Sreerup Raychaudhuri
May 2021

Contents

Part I
The Coming and Spread of Western Science

Prologue

Abu-Rayhan Muhammad ibn-Ahmad al-Biruni (973–1050) was a famed Muslim scholar from Khwarizm in Central Asia, who, after various vicissitudes, settled at the court of the intrepid warlord Sultan Mah'mud of Ghazni. This potentate made it his business to lead annual raids into India (1001—26), and al-Biruni accompanied him on these plundering expeditions. However, while the Sultan collected mounds of looted gold and slew 'infidels', al-Biruni was more interested in garnering all the knowledge he could from the 'Hindu scholars', as he somewhat naïvely described the Brahmin pundits living in the plains of the great river Sindhu—known to the Persians as 'Hindu'. Al-Biruni left an extensive documentation of what he called *'the reasonable and the unreasonable'* beliefs of these alien savants, translated some of their works into Persian and quietly passed off some of their ideas as his own. Paradoxically, these naïve writings now form the best picture we have of Indian society at the beginning of the second millennium of the Common Era.

Gathering information from the xenophobic Brahmins was not easy. Even though independent thought in India had already been in decline for a few centuries, al-Biruni complained that *'The Hindus believe that there is no country but theirs, no nation like theirs, no king like theirs, no religion like theirs, no science like theirs.'* While such conceit is true of many a land, the statement was largely true of Indian science at that time, not least because the ritual Hindu prohibition on foreign travel automatically created an insular outlook. With somewhat rueful honesty, however, al-Biruni also admitted that *'Hindu sciences have retired far away from those parts of the country conquered by us, and have fled to places which our hand cannot yet reach…'*.

In time, al-Biruni and his fierce patron died. Within some more decades, the power of the Ghazni Sultanate waned and was replaced by that of the warlords of Ghur, another fortress town, now in modern Afghanistan. Around 1192–1995, the ruler of Ghur, one Shahab-ud-din, not content with mere raids and plunder, conquered northern India. One of his skilled generals, who rejoiced in the name of Ikht'yar-ud-din Muhammad bin-Bakht'yar Khilji, carried fire and sword through eastern India. In the process, he launched an attack on the great Nalanda *Mahavihara*—the famed

Buddhist monastery-cum-university. This happened in or around the year 1194. The historian Minhaj us-Siraj, writing a half-century or so later, chronicles that the Khilji warlord and his soldiers were rather taken aback when the shaven-headed 'defenders of the fortress' sallied forth without arms, chanting some unfamiliar 'war-cries'—but decided to slaughter them anyway. This little footnote in the history of the Khilji's sanguineous expedition is generally considered to be the point at which India slipped into her Dark Ages.

Of course, this dramatic incident was merely the *coup de grace* to a culture that was already in decline. In fact, five centuries were to pass during which Indian society—Hindu and Muslim—degenerated apace and gradually sank deeper and deeper into the mire of superstition, losing all contact with science. In one of the great ironies of history, this happened at the very same time that the translations of Indian manuscripts by al-Biruni and his successors were setting off a tremendous efflorescence in the Mohammedan world in the selfsame fields of chemistry, mathematics and astronomy. Some of this, through Latin translations of the works of al-Hay'tham (Alhazen), ibn-Sina (Avicenna) and ibn-Rashed (Averrhoes), reached the Western world and lay there ready for the Renaissance. The Renaissance, as is well known, was finally triggered by the fall of Constantinople (1453) and the consequent flight of Greek scholars to the West, leading to the rediscovery in Western Europe of the ancient Greek traditions of free thought. Western science then commenced to pull itself out of its own Dark Ages—and has never looked back.

During the intervening period, however, 'the Great Mogul' ruled India in splendid opulence, while the common people remained in the clutches of darkest superstition. In particular, Hindu society—for by this time 'Hindu' had come to generically denote the clutch of original religions and sects indigent to India—divided itself rigidly into castes and sub-castes, and developed several ghastly abuses, of which the burning alive of widows (*sati*) forms just one example. However, it is not as if the Islamic section of society was more enlightened. Apart from the constant stream of coarse military adventurers who poured in from Central Asia to seek their fortune at the courts of the Emperors, the bulk of Indian Muslims were converts, often from the unlettered lower classes of Hindu society. Polygamy, total segregation of women (*purdah*) and fratricidal conflict among princes were the hallmarks of Islamic society in India's Dark Ages. Irrespective of region, religion or social status, the 'sciences' of this period were alchemy, astrology and black magic.

The Europeans who came to India towards the end of the Mughal period were hardly out of their own Dark Ages. The first to come were the Portuguese and the Spanish, followers of the resurgent Catholic faith which had produced the Inquisition. Fresh from having subjugated and decimated the indigenous peoples of the Americas, these traders—for it was the spice trade which lured them to make the long and dangerous sea voyage to India—were no less ruthless or iconoclastic than the Turkic conquerors who had burst into north India three centuries before. The remains of broken temples and defaced idols are spread all along India's coastal regions as mute witness to their pious fervour.

In the wake of the Portuguese came the Dutch and the Danes and the French and the English—all intrepid seafaring nations, unwilling to allow the militarily weaker

Portuguese to monopolise the spice trade from India. The biggest prize went to the English. Residents of a small island, they ended up ruling a vast territory 20 times larger, comprising the five modern states of India, Pakistan, Bangladesh, Sri Lanka (Ceylon) and Myanmar (Burma). The story of this conquest does not concern us here. Suffice it to say that once the British *raj* (i.e. rule) became stable, even as India was subjected to the systematised loot which was mercantilism, European ways and European thought got stirred into the great melting pot of Indian culture. This changed India forever, pulling the great subcontinent willy-nilly out of the mediaeval age into the modern one.

Western science in India was not, therefore, an indigenous growth, but was grafted and nurtured like an exotic plant by foreign rulers, though it eventually came to flourish without external props. In fact, when the Europeans came to India as traders, and to everyone's surprise—including their own—ended up ruling the country, there was tremendous resistance to the scientific ideas they brought with them. For Hindus and Muslims alike thought of these concepts as intimately associated with Christianity. So indeed, thought the more aggressive proselytisers among the Europeans. In fact, however, we can identify *three* major factors which largely contributed to arousing Indian interest in Western science and technology.

The first reason was undoubtedly political. By the middle of the nineteenth century, the British *raj* was fairly well established all over the geographical entity nowadays referred to as South Asia. This runaway political success of what was initially a merchant company naturally led native-born pundits to start examining its causes— and they found it convenient to attribute it to the benefits of Western science and a 'more materialistic' attitude towards life. It was comforting to forget the greed, mutual hatred and sheer incompetence of India's native potentates and attribute everything to the legend of superior British firepower.

Another stimulus was economic. No one could fail to notice the impoverishment and ruin of the traditional village-based industries of India and their replacement by large-scale imports of machine-made goods from the burgeoning factories of industrial-age Britain. Much of this was, in fact, due to ruinous tariffs imposed by the conquerors on native industry, but to the eye of the consumer, Western technology was clearly winning this round against Indian traditional manufacture. It made sense, then, for the subalterns to get into a business which promised both wealth and progress.

A third thread which arose from the colonial dispensation was evangelism. On the side of the conquerors, interwoven with ideas from the Enlightenment, was the underlying hope was that if 'the heathen could be weaned away from their dark superstitions' by teaching them science and the scientific temper, then, in no time they would come readily to the waiting arms of Jesus. In this matter, however, the evangelicals badly underestimated the strength and resilience of the 'native' ways of thought. India did not turn into a Christian country, but into the curious hotchpotch of East and West which it is today.

The more down-to-earth rulers of British India were, however, well aware of the fact that their hegemony would last only so long as the 'natives' viewed them with awe as a superior race with superior skills. Initially, they kept their knowledge of science strictly to themselves. Nevertheless, European ideas were quickly absorbed by the so-called *bhadralok* (gentleman) class, which initially consisted of Bengali upper-caste Hindus who quickly adapted to the novel concepts of Enlightenment Europe. Soon the *bhadralok* spread across India and across castes, creeds and religions, and they all clamoured to learn science. It was in this atmosphere of British parsimony versus Indian aspiration that Western science made its modest beginnings in the great subcontinent.

Chapter 1
The First Steps

1.1 Map-Making in the Fledgling *Raj*

In 1685, an enterprising English merchant called Job Charnock (1630–92) became the head of the British East India Company's interests in Bengal. Bengal was then the richest province of the mighty Mughal empire, which held sway over most of India. Charnock decided to move his centre of operations from Cossimbazar (Qasem Bazaar), which was rather too close to the Mughal Governor's seat of power, and move down the river. In 1690, the East India Company managed to obtain an imperial *firman* or decree which permitted them to trade free of duty in Bengal and to build for themselves a fortified settlement. Charnock chose the village of Kalikata as the site for this fort, and named it Fort William. Around this site grew the city of Calcutta (Kolkata).

Charnock died in 1692, but the settlement flourished, while the 'factors'—as the men who ran factories were called in those days—took good care to keep on the right side of the Mughal Governor. After the death of Aurangzeb's grandson in 1713, Bengal became effectively independent under her Nawab, or Governor. In 1756 came a new Nawab—Siraj ud-Daulla—young, arrogant, naïve and rapacious. In a few months, Siraj had succeeded in alienating all around him and thus a conspiracy was formed to eliminate him and replace him with his more amenable general Mir Jaffar. It was at this juncture that the Nawab decided to claim from the English at Calcutta the huge arrears of customs duties which had not been paid over the preceding decade or two. The English refused, and war ensued. The British were rallied by a clerk-turned-soldier named Robert Clive. Under Clive's able leadership the Nawab was defeated and killed. The erstwhile general Mir Jaffar was made the new Nawab under the sonorous title *Syed* Mir Jaffar Ali Khan *Bahadur*.

Clive prevailed upon the pliable Nawab to allot to the East India Company a large tract of land around Calcutta corresponding to twenty-four *parganahs*—a *parganah* being the Mughal equivalent of a subdivision—from which the Company alone would draw taxes without requiring to pay anything to the Nawab. This area later became

© The Author(s), under exclusive license to Springer Nature Switzerland AG 2021
S. Raychaudhuri, *The Roots and Development of Particle Physics in India*,
SpringerBriefs in History of Science and Technology,
https://doi.org/10.1007/978-3-030-80306-3_1

the British-Indian district of 24 Parganahs, and remains an administrative unit until the present day.[1]

It was at this tumultuous moment in Indian history that Western science made its first tentative entry to the great subcontinent. Clive's successor as the head of the East India Company in Bengal was Henry Vansittart (1732–70). Taking charge in 1760, he found the Company's finances in complete disarray and resolved that every last coin of revenue would be extracted from the 24 Parganahs, where tax collection was lax and evasion was common. For this, it was necessary to maintain proper records, with maps and measurements.

Enter James Rennell (1742–1830). A Devonshire boy, trained as a seaman, he had a good head on his shoulders and quickly learned to make marine charts and take bearings, as every good navigator should. He made ample use of these skills in the Royal Navy, which posted him to Madras. In 1763, he decided to quit and join the East India Company instead. He was posted to Bengal in 1764, where Vansittart commissioned the youth to survey and create a 'Domesday Book' for the 24 Parganahs. So well did this seaman-turned-cartographer do the job, that Vansittart appointed the 21-year old to the post of Surveyor of Bengal and gave him the job of mapping out the whole of the Bengal province, which he did. The Directors of the Company eventually recalled Vansittart and sent back Clive for a second stint in India. In this second innings, Clive gave to James Rennell the grandiose title of Surveyor-General of India and commissioned him to map the entire plain of the river Ganges, up to Benaras (now Varanasi). All of this was now essentially British territory. Rennell went further, as the British domains increased, and ended up creating the first scale maps of northern India—the first time anyone had done so using the proper concept of a geometric projection and a scale.

The iniquities of Robert Clive and his successor Warren Hastings in India alerted the British Government to the urgent need to combine the Company's power with responsibility. And so, in 1773 came a Regulating Act, and in 1784 Pitt's India Act. The British in India were no longer to be adventurers and freebooters. The emphasis changed to exploiting India's natural resources rather than looting her people. This was a task that required scientists, engineers and explorers to find out where the riches of the land lay, prior to taking them out and shipping them off to the British isles. And this is how—and why—Western science first came to the subcontinent.

Rennell continued with his mapmaking activities, basing himself at Dacca (Dhaka in modern Bangladesh), and eventually mapped out the whole of north India, from the Himalayas in the north to the jungles of Bundelkhand in central India, and from Agra in the west to the river Meghna in the east. In 1776, however, James Rennell and his surveying team fell in with a party of rebels who set upon them and beat them within an inch of their lives. Rennell survived indeed, but he lay in a Calcutta hospital for almost a year and then left for England, shaking the dust of India off his feet for ever. Thus ended the innings of the first Western scientist to live and work in India.

---

[1] In recent times it has been divided into a North and a South district.

1.2 The Great Trigonometric Survey

While Clive and Hastings consolidated their rule over Bengal, a new power had arisen in the south. By the early 1760's, the kingdom of Mysore was completely controlled by an adventurer named Hyder Ali. He was succeeded by his son, Tipu Sultan, who attempted to consolidate the kingdom which his father had usurped, but did not have time to settle. The British attacked him and Tipu was brought to bay by the combined forces and besieged in his capital of Seringapatam (Srirangapattana). Here, when the defences were finally stormed, he scorned to flee and instead fought and died a soldier's death in the breach.

Among the British commanders at the siege of Seringapatam was a young nobleman called Arthur Wellesley. This officer was the younger brother of Lord Wellesley, then Governor-General of India, and he was destined for a great future as the Duke of Wellington. Fighting under Colonel Wellesley's command was a middle-aged lieutenant called William Lambton (1753–1823), who had displayed conspicuous perspicacity in the manoeuvres preceding the storming of Seringapatam. A Yorkshire man, Lambton came from a humble background and had sought his fortune by joining the British army. He was already a seasoned officer by the time he was posted to India.

Lambton also had a bent for mathematics, and had taught himself the science of trigonometric surveying and geodesy. During the Seringapatam campaign, he became convinced of the need for accurate maps to conduct warfare in a foreign country, where the enemy would be native people familiar with the lie of the land. And so, in the year 1800, he wrote a proposal to perform a proper scientific survey and map out the entire south Indian peninsula region, creating topographical scale maps just as the late William Roy (d. 1790) had done for England, Scotland and France. This proposal he sent to his chief, Colonel Wellesley, who enthusiastically forwarded it to Lord Edward Clive,[2] the Governor of Madras, for his approval and for the money required to carry the work out. A recommendation from such a source could not be ignored, and so the Governor sent off the proposal to England, to be vetted by none other than James Rennell, who was then the acknowledged expert on all things cartographic.

At age 58, Rennell was now an old fogey by the standards of the time, and he did not think highly of the method proposed by Lambton. The basic method used in all surveying is that of triangulation. The idea is that if three points A, B and C form a triangle, then, by measuring the length of the base BC and the two angles on either side, i.e. $\widehat{ABC}$ and $\widehat{ACB}$, one can determine the distances AB and AC by using a simple trigonometric formula (see Figure). One then uses one of these sides as the base for the next triangle, and so on, until there is a network of points whose precise positions are known. The map is then drawn around this network of fixed points. If the triangles in question are small compared to the size of the Earth—as they usually are—it is a good approximation to consider them as plane triangles,

---

[2] Son of the great Robert Clive.

and use the simple formulae of plane trigonometry. This was, in fact, what Rennell himself had done, in keeping with the general practice of his day. However, the Earth's surface is really curved, and so this method introduces tiny errors every time the plane trigonometric formulae are used. As the results from each triangle are used to measure the next, these errors keep building up. Roy's method, which Lambton proposed to follow, was to use *spherical* trigonometry, which would eliminate these errors and give a much more accurate measurement of distances. However, like many a senior scientist, Rennell felt that all this was unnecessary sophistry, and that the good old methods were adequate for the job. And so he wrote back that Lambton's proposal should be rejected.

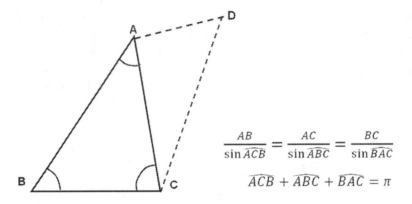

$$\frac{AB}{\sin \widehat{ACB}} = \frac{AC}{\sin \widehat{ABC}} = \frac{BC}{\sin \widehat{BAC}}$$

$$\widehat{ACB} + \widehat{ABC} + \widehat{BAC} = \pi$$

Governor Clive was now in a quandary. The issue seemed to be one of mathematics, so with great discernment, Clive sent it out again to the Astronomer Royal of England, the Rev. Dr. Nevil Maskelyne, who, apart from his interest in the skies, was a genuine expert on geodesy. Reading Lambton's proposal, Maskelyne was immediately struck by its merit. Perhaps the part which appealed to him most was its purely scientific component. For one of the hot topics in geodesy then was to see if the Earth is indeed a sphere, or shows small deviations from sphericity (it does!). The simple way to do that was to measure one degree of latitude along the surface of the Earth in different places and see if they matched. Any mismatch would point to a deviation from perfect sphericity. Several such measurements had been made in Europe and the New World, but Lambton was the first to propose a measurement in the tropical latitudes. This may have excited Maskelyne, for we know that he dashed off a letter to Madras mentioning his enthusiastic support. Lord Clive now felt confident to approve Lambton's proposal and assign it enough money. And thus was born the Great Trigonometrical Survey, one of the finest scientific enterprises of the nineteenth century.

Lord Clive had given a sanction for a five year project to map the lands under British rule in south India, but the project kept getting extended and continued for sixty years, eventually spanning the entire length and breadth of the Indian subcontinent. The beginnings were simple enough. For equipment, Lambton was lucky to find a high quality zenith circle (a primitive theodolite) and a surveyor's chain in

Calcutta. These had not been brought there to help survey Bengal, or any place in India for that matter. Instead, they had formed part of the abortive Macartney mission to China (1792–94). The instruments, a bit rusted, but quite usable after some spit and polish, eventually reached Lambton at Madras, and enabled him to get the Survey going.

The Great Trigonometric Survey began from the famed St. Thomas' Mount, where the Apostle Thomas is reputed to have suffered his martyrdom, and which is now integrated into the bustling metropolis which is modern Chennai (formerly Madras). From there, a 'cobweb' of triangles spread out to cover all of peninsular India, producing scale maps of an accuracy hitherto unachieved in the whole of South Asia. Year by year the grid spread northward, entering the dense jungles of Central India, where Lambton and his team soldiered on amid terrible dangers. The jungles of Vidarbha were often so dense that it would take weeks to clear an opening where the surveying marks could be sighted. Sudden thunderstorms would blow away their tents and send their sensitive equipment crashing to the ground. There were prowling tigers which would come silently in the night and take away one of the coolies, deadly snakes that could kill a strong man within the hour, venomous scorpions by the dozen and a thousand stinging insects, including clouds of mosquitoes. In fact, the last almost put paid to the Great Trigonometric Survey, for every single member came down with attacks of malarial fever.

Malaria was treated at the time with powdered bark of the cinchona tree, mixed with wine, which is good enough to effect a cure, but medical practitioners of the time thought it should be accompanied by a diet of bread and water, frequent bleeding and strong purgatives to 'take out the corrupt humours'. This treatment often proved more deadly to the weakened patient than the disease itself. Such a fate overtook Lambton himself, for despite his iron constitution, he fell ill in 1823, somewhere in the heart of the Deccan plateau, and suffered the doctor on his team to give him the customary treatment. This helped sped him to his unremarkable grave at the unremarkable town of Hinganghat. But it is perhaps fitting that a man who devoted his life to geometry should spend eternity at almost the exact centre of the Indian subcontinent.

1.3 The Cantankerous Surveyor

With Lambton permanently under the ground, the task of carrying on 'the greatest scientific enterprise of the nineteenth century' fell upon a man who was of a very different stamp, but who shared Lambton's single-minded passion for getting the work done. This was George Everest,[3] later to become Sir George Everest (1790–1866). Of Welsh extraction, Everest was born in Greenwich, almost under the shadow of the Royal Observatory. As a youth, he was commissioned into the artillery, and at

---

[3] Pronounced 'Eev-rest', with the 'Eev' rhyming with 'leave'. Sir George was very touchy about this, as with many other things. However, the mountain named after him is usually known as 'Ever-rest', with the 'Ever' rhyming with 'Never'.

age 16, he was made a Lieutenant, and sent off to India, and thence to Java. In 1814, he was posted at Calcutta, having become a confirmed expert in surveying, and was tasked with carrying out a survey to find the best route for a semaphore telegraph line from Calcutta to Benaras, which he again completed in 1816–17. Some of those semaphore towers still stand, and were, in fact used later as triangulation points by the Survey.

By 1818, Lambton, still in the jungles of the Hyderabad state, triangulating away in company with the half-Indian assistants trained by himself—Joseph Olliver, William Rossenrode, Joshua de Penning—had heard of the bright young surveyor, and immediately wrote to the authorities at Calcutta, asking that he be transferred to the Survey. His wish was granted, and so the young officer was transferred from the fleshpots of Calcutta to the jungles of Central India. Recognising his outstanding abilities, Lambton immediately made Everest his deputy, putting him over the heads of the older and more seasoned lieutenants he had with him. Thus, when Lambton died in 1823, it was natural that the 33 year old Everest would be named Superintendant of the Great Trigonometric Survey. The remaining story of the Great Survey is an unexciting tale of dogged and dedicated toil in the face of enormous odds. The unhealthy climate of Central India nearly destroyed the project, for in 1825, Everest—who did not have Lambton's iron constitution—fell too sick to continue and was allowed to take ship for England to convalesce.

Back in England, Everest slowly regained his strength, while the work of the Survey was carried out, extending the web of triangles towards the east, where Calcutta was, by the able lieutenants trained by Lambton. By 1830, Everest was back in India, with his health recovered, but not his temper, for he now developed a habit of hysterically berating his subordinates at the slightest hint of real or imagined error. So outrageous was his behaviour that popular historian John Keay describes him as '*the most cantankerous* sahib' British India had ever seen. Fortunately, Everest refused to venture into the jungles again, and confined his scolding to nasty letters. He himself sought the cooler climate of Dehra Dun, where the Survey of India still has its headquarters. For indeed, the erstwhile Lieutenant Everest had returned to India holding the superior rank of Surveyor-General of India and Superintendant of the Great Trigonometric Survey. Meanwhile, Joshua de Penning had been sent by Everest to take over the Calcutta office of the Survey, which now became a centre for the Survey's 'computers', i.e. mathematically-trained clerks who would make all the painstaking calculations based on the raw data noted down by the surveyors.

It would be 1866 before the Great Trigonometric Survey could be declared completed. Everest himself had returned to England in 1843 and would live to see his life's ambition completed, for he died in that very year 1866, aged 76. He had been made a Colonel in 1854, and was knighted in 1861. Sadly, there had been no such recognition for William Lambton, the founder of the Survey, a more genial person, and arguably, a more original scientist than his more famous successor.

Chapter 2
The Highest Mountain in the World

2.1 The Himalayas

What happens when an irresistible force meets an immovable obstacle? The answer is—the Himalayas. Some 55 million years ago, a fast-moving tectonic plate, pushed northwards by a bigger plate containing the vast continent of Africa, crashed into the even bigger static plate of Eurasia. This moving plate, which we call the Indian plate, had a northern half which was ocean, and a southern half which was landmass. For another 20 million years, the thinner oceanic crust of the Indian plate slipped under the Eurasian plate, causing a portion of it to rise up exactly like soil under a spade. This is what we today call the Tibetan Plateau, or, in more poetic language, The Roof of the World. But following behind came the thicker crust carrying the landmass, and when this hit the southern edge of the Tibetan Plateau, it was too thick to go under. Instead, the edges of both the plates rose up sharply, forming a long and high mountain range. These are the Himalayas, without question the highest mountain range in the world. This collision process is still under way. In fact, the Himalayas are still growing at the rate of about a centimetre per year—agonisingly slow for us short-lived creatures, but simply galloping on the geological time scale.

As Egypt is the gift of the Nile, so India is the gift of the Himalayas. Without the high wall of the Himalayas, the freezing winds from Siberia and the Arctic would have swept over the subcontinent, chilling it to subzero temperatures most of the year. Likewise, the moisture-laden monsoon winds which sweep in from the southern ocean would have crossed the low hills and vast plains of India and deposited their life-giving rains in the uplands of Tibet. Had there been no Himalayas, the interior of the Indian subcontinent would have been a howling desert like the Gobi or the Takla Makan, or, at best, a freezing steppe land like the vast plains of Central Asia. The fertile fields and lush vegetation of India, her dense tropical jungles and broad rivers, the teeming population and a classical image of being a land of milk and honey, are all due to this mighty barrier of tall mountains which stands sentinel over the northern boundary.

© The Author(s), under exclusive license to Springer Nature Switzerland AG 2021   11
S. Raychaudhuri, *The Roots and Development of Particle Physics in India*,
SpringerBriefs in History of Science and Technology,
https://doi.org/10.1007/978-3-030-80306-3_2

To Indians, through the ages, the Himalayas have been sacred mountains, the abode of the gods and the source of a dozen holy rivers including the revered Ganges. But it was not in the South Asian psyche to believe in sketching or mapping the vast mountain range, which stretches in a great arc about 2500 kms in length and covering an area of nearly 600,000 square kilometres. They were simply content that it was there. In the eighteenth century, however, the great French cartographer, d'Anville (1697–1782) completed his celebrated map of China with a map of Tibet and a first-ever scale map of the Himalayas. It was conjectured by D'Anville that the Himalayas were much higher than the Alps, but in those days the Andes were believed to be the highest mountains in the world.

Once British control was firmly established over the area, a few officers began to journey into the hills, which they found delightfully cool during the searingly hot months of the Indian summer. It is interesting that, though the irascible Surveyor-General of India enjoyed the cool climate of the mountains and no doubt admired the view of the snow peaks from nearby Mussoorie and Chakauri, his scientific interest in these white-capped giants was rather limited. Andrew Scott Waugh (1810–78), the Scotsman who became Surveyor-General when the 'old man' retired in 1843, was different. An army colonel like Everest and Lambton before him, he too ended his life as a knight and a Fellow of the Royal Society. In those days, however, his main passion lay in surveying the far-flung Himalayan range and measuring the height of its towering peaks.

2.2 Measuring the Great Peaks

Even before he became Surveyor-General, Waugh had begun to indulge his passion for surveying the mountains. Enlisting the young Charles Lane, a bright young surveyor, well-versed in the mathematical techniques of the trade, he first carried out measurements of the high peaks visible from the towns of Uttarakhand, the hilly region north of where Dehra Dun lies. For a long time, the highest mountain in the world had been presumed to be Mt. Chimborazo, a towering volcanic peak in the Andes, measuring 20,548 ft. above mean sea level. By the turn of the nine-teenth century, however, British surveyors[1] had found a Himalayan peak—Mt. Nanda Devi—which, at 25,643 ft., is higher than anything the Andes have to offer. In 1808, however, the neighbouring peak of Dhaulagiri, standing at 26,795 ft. was found to be still taller. For the next 39 years, it would be regarded as the highest in the world.

---

[1] Andrew Scott Waugh was not the first British surveyor to be fascinated by the Himalayas. The first to explore the lower ranges and measure some of the peaks was Captain William Webb, a onetime associate of Colebrooke, who had been given the responsibility of surveying the area of Kumaon, now in the state of Uttarakhand. Another officer, Captain John Hodgson, was to survey the area of Garhwal. These two districts had been snatched away from the Gorkhas in the war of 1814–15, and it was in 1816 that the surveys started. Hodgson was joined in 1819 by James Herbert, and together they measured the peak 'A2' — Mt. Nanda Devi — to be 25,749 ft. above sea level, and consequently the highest known peak. They were accurate to within a hundred feet.

The Himalayas are not confined, though, to the Uttarakhand region, but sweep down in a great arc all the way to the edges of the Shan Plateau some 2,000 km away, and then bend south through Myanmar in a cordillera known as the Arakan Yoma. East of Uttarakhand, where the Himalayas form the southern wall of the Tibetan plateau, lie a succession of tall snow peaks, which can rival the giants of Uttarakhand, and cried out to Waugh for measurement. However, apart from the handful of peaks in the Uttarakhand region around Dehra Dun, the great peaks of the Himalayas all lay in forbidden territory. In the north, Tibet had closed her borders to foreigners at the conclusion of the drawn war with Gorkha-ruled Nepal in 1792. In Nepal itself, where the bulk of the high peaks lay, Englishmen and their satellites were *persona non grata* ever since an Anglo-Gorkha war had ended in 1816. The far-western end of the Himalayas belonged to the Khalsa kingdom of the Sikhs, where an uneasy peace existed in the aftermath of the first Anglo-Sikh war (1845–46). The eastern end of the mountains, which are somewhat less high, belonged vaguely to China and to a Burma (Myanmar) which was licking its wounds after the first Anglo-Burmese war (1824–26). And so, in the 1840s, most of the Himalayan region would probably have meant death or a dismal imprisonment for any British expedition which trespassed into non-British territory.

There was thus only one good vantage point from which Waugh and his merry men could see the great peaks lying all along the Nepal-Tibet border, and that was from the hills in the north of Bengal, in the neighbourhood of Darjeeling. Here a series of 'hill stations' came up one after the other—Darjeeling, Kurseong, Kalimpong. From these towns, one only has to look north to see a towering range of mountains, which is locally called *kang-tsen-dzo-nga,* which, in Tibetan, means 'five treasures of snow' referring to the five great peaks which can be easily distinguished as taller than all the rest. The highest of these is a sharp pyramidal peak, which Waugh and his assistant Charles Lane measured in 1847 as having a height of 28,156 ft., considerably higher than Mt. Dhaulagiri at 26,795 ft. As a result, Waugh was able to announce that this peak—which he called by its Bengali version 'Kanchanjungha' (meaning 'golden thigh')—was the highest known till date.

From their triangulation point at Sonakhoda in north Bengal, however, Waugh and Lane could look north-west, and observe the mighty palisade of the Himalayan range in the forbidden region where Nepal borders on Tibet. And there, towering distinctly above the 20,000 ft. wall of ice were three distinct spires, all of which seemed as tall, if not taller, than Kangchenjunga. Of these, the central one—which Waugh rather unimaginatively called γ—seemed to stand a bit above the other two, though it was difficult to say for sure from about 230 kms away. Waugh wanted to know their height, for he had a sneaking suspicion that one or even all of these might rival Kangchenjunga. However, he had completed the job assigned by Everest, and his party were almost prostrate with 'jungle fever' i.e. malaria. It was time, therefore, to pack up their equipment and return to the more salubrious climate of Dehra Dun. The question about the three mysterious peaks remained, however, as a niggling disquiet in the mind of the dogged Scotsman.

Waugh and his team were not the only ones to spot these three towering peaks. They were independently spotted by John Armstrong, another member of Waugh's team, who was working from Muzaffarpur, in the north of the present Indian state of Bihar, about 50 kms from the Nepal border. Armstrong had named these giants as Peak a, Peak b and Peak c. His result, for Peak b was 28,799 ft. which was considerably higher than Kangchenjunga. However, Armstrong also found a similar height for Peak a. When Armstrong submitted his report to Waugh at Dehradun, the latter guessed that this Peak b was probably the same as his own γ, but was naturally suspicious of the claim of *two* mountains taller than Kangchenjunga. In 1849, therefore Waugh, now Surveyor-General of India, reorganised the Himalayan survey on a war footing. To begin with, he tasked a young assistant, John Hennessey, barely 20 years old at the time, the job of renaming the Himalayan peaks in a systematic fashion. Hennessey did this admirably, using Roman numerals, and so the unknown Peak b of Armstrong became Peak XV. At the same time, Waugh sent James Nicholson, a more experienced man, to check on Armstrong's work. Nicholson complied, and managed to take detailed theodolite sightings from six different stations along the India-Nepal border. Retiring to Monghyr during the rains, when field work was impossible, he then did a preliminary 'field computation' and found Peak XV to be 30,200 ft. This was the highest peak with a vengeance, but it made Nicholson's boss even more suspicious. And this made him send off the whole set of Nicholson's data to his best mathematician, whom he had posted earlier in 1849 to Calcutta as Chief Computor of the Survey of India. This man was Radhanath Sickdhar,[2] and it is because of him that this story makes its way into the present work.

2.3   The 'Hindoo Savant'

Radhanath Sickdhar (or Sikdar), who can lay undisputed claim to the title of being the first modern scientist from India, was a product of the Hindu *Mahapathshala* (later Presidency College) which will be described in the next chapter. A star pupil of the Mathematics Department, he was a special protégé of the Professor of Mathematics, John Tytler. Sickdhar had joined the Survey of India in 1831. And therein lies a tale, which should be taken up from the beginning.

Radhanath was born in 1812 at Sikdarpara, a part of Jorasanko in north Calcutta. His father, Tituram Sikdar, was determined that his sons would be educated in the *firangi* style and become clerks in some merchant company. At the precocious age of 11, therefore, Radhanath entered the exciting environment of the Hindu *Mahap-athshala,* or Hindoo College as the Britishers called it. Here he was introduced to a host of new subjects and new ideas, as most undergraduates are, but more importantly, he fell under the spell of Derozio.[3] This was to prove the defining influence

---

[2] The modern spelling is Sikdar, but Sickdhar was Radhanath's own spelling and has been used throughout this work.

[3] Employed as a professor of English at age 17, the larger-than-life poet Henry Louis Vivian Derozio was a radical reformer who gathered around him a band of admiring students whom he taught to question faith and superstition. Dismissed after five years for 'corrupting the youth', he died very

in his life. All the while, however, Radhanath soaked up knowledge like a veritable sponge, though it was in mathematics that he found his true love. In 1830, having passed his undergraduate degree, he began to read advanced mathematics with Professor John Tytler, their revered mathematics teacher. Between 1828 and 1832, Radhanath mastered Euclid's *Elements*, Windhouse's *Analytical Geometry* and *Astronomy*, Jephson's *Fluxions*, and finally Newton's massive *Principia*. His studies of Euclidean geometry inspired him, while still an undergraduate, to discover a new way to draw a common tangent to two circles, a problem which had been discussed a myriad times since Euclid. But Radhanath's interest in geometry was not confined to plane geometry. He was fascinated by spherical geometry and spherical trigonometry, perhaps inspired by the intricacies of the *Principia*. This was soon to become his life's preoccupation.

While Radhanath was navigating the mazes of Newton's *Principia*, George Everest came knocking at the doors of the Hindoo College. The Governor-General, Lord Bentinck—whose main mission in India was cost cutting—had, just a year before, in 1829, written in a minute on the Survey that "*It is by a more enlarged employment of native agency that the business of a Government will be at once more cheaply and efficiently transacted.*" And so he advised Everest to hire 'native' boys who had some mathematical training and teach them to do the grunge work required for the Survey. It would still pay them more what they could get as lowly clerks in a merchant company or a *zamindar*'s office. The roots of 'outsourcing' do, indeed, go deep!

Having no alternative, Everest proceeded to go to the only place where he might find 'native' boys with the requisite training, and that was to Professor Tytler at the Hindoo College. In response to the Surveyor-General's letter, Tytler wrote back that that he had the very boy whom the Survey was searching for, and that was young Radhanath Sickdhar. And so, in 1831, at the age of 19 years, the first Indian scientist of modern times was employed by the Survey. Radhanath displayed such aptitude for the work that, by the end of 1832, Everest had carted him off to his headquarters at Dehradun, there to work closely with himself in the great triangulation project for north and west India. To old Tituram Sikdar, Everest wrote, in somewhat quaint terms, "*I wish I could have persuaded you to come to Dehra Dun for not only would it have given me the greatest pleasure to see you personally how much I honour you for having such a son as Radhanath, but you would yourself have, I am sure, been infinitely gratified at witnessing the high esteem in which he is held by his superiors and equals*". This was high praise, indeed, from the old curmudgeon, even if his sentence construction was a bit awry.

Like his predecessor, Waugh was highly appreciative of the talents of the 'Hindoo savant', and this led him to take an unprecedented step. In early 1849, he posted the talented Bengali to the Survey's office in Calcutta as Chief Computor. This was the first time an Indian had been appointed by a Britisher to a post under the East India Company where there would be Europeans working as his subordinates, and

---

soon after, aged but 22. During this short life he may be said to have fairly set off the Bengal Renaissance.

was a mark of the great impression made by Sickdhar on his mentors and superiors. Waugh knew, in particular, that this prodigiously talented theorist had made original contributions in the art of making corrections to raw surveying data by taking into account unavoidable error-inducing factors like atmospheric refraction, the expansion and contraction with temperature of the measuring chains, the curvature of the Earth, and so on. He was an authority on the data crunching required for accurate surveying and an early advocate of the now-ubiquitous least-squares method of error minimisation. In fact, he was the right man in the right place.

2.4   The Roof of the World

Receiving the data from Nicholson's surveys in 1850, Radhanath and his associates toiled over them for two years, checking and re-checking their own calculations, and applying all possible corrections in a bid to determine the correct height of the mysterious Peak XV. By 1852, they were ready to announce their results. The six independent measurements made by Nicholson yielded:

Place	Jarol	Ladnia	Mirzapur	Harpur	Janjapati	Menai
Average (ft)	28,992	28,999	29,005	29,026	29,002	28,990

Together, these yielded an average height of 29,002 ft., with a standard deviation of 140 ft. It is a tribute to the quality of measurements made by Nicholson and his team, no less than the computations made by Radhanath and his team, that the error was barely half a percent.[4] There was no question now that Peak XV is some 800-odd feet higher than Kangchenjunga and therefore the highest mountain in the world. Nevertheless, Andrew Scott Waugh was a model of scientific caution. Suppressing the great excitement he must surely have felt, he handed over all the data to an independent team at Dehradun under the supervision of John Hennessey, and waited four long years till 1856, by which time the Dehradun team had confirmed the results of the Calcutta team *in toto*. Waugh now felt confident to let his assistant Captain Henry Thuillier announce at the August 1856 meeting of the Asiatic Society in Calcutta that[5] "... *the final values for the peak designated XV of the Trig. Survey, and which place it in N. Latitude 27° 59′ 16″-7 and 86° 58′ 5″-9 Longitude E. Of Greenwich, with an elevation of 29,002 feet above the sea level, or 846 feet above Kanchinjinga, and 2,176 in excess of the far famed Dewalagiri.*" Thuillier, reading

---

[4] There is an urban myth that the Chief Computor and his team (or Hennessey in other accounts) found the height of Peak XV to be exactly 29,000 ft., but their boss, Andrew Waugh, told them to put on another 2 ft. in order to make the result look plausible. The origin of this story is not known, but whoever made it up had clearly not seen the table of results nor did he know of the upright character of Waugh.

[5] The original spellings and punctuation from the *Proceedings of the Asiatic Society* for 1856 are reproduced in this quotation, including the hyphens in place of the decimal points.

from Waugh's letter, went on to announce that he (Waugh) could find no local name assigned to this lofty pinnacle, and hence, proposed that it should be named 'Mont Everest'. A year later, in a formal letter written to the Royal Geographical Society, Waugh changed the name to the more conventional Mount Everest, and that is the name by which we know it today.

The great discovery had been made. It was a triumph of science and perseverance and belongs forever to the glorious history of humankind. The rest of the story is also human, for the 'roof of the world' has had its own fair share of controversies, and not a little bit of myth-making. In fact, one just has to type 'Mt. Everest' on the Internet to be confronted by a perfect storm of cant and misinformation. It was only after a decade of controversy that the Royal Geographical Society accepted Waugh's proposed name in 1865. Curiously, Sir George Everest died within a year of his name becoming immortal. However, the controversy was not laid to rest with him. In the twentieth century, Sven Hedin, the Swedish explorer, visited Tibet and discovered a longstanding Tibetan name, viz. *Chomolungma*. Since then, this claim has been propounded belligerently and incessantly by the Chinese. It seems, however, that the Tibetans refer to the entire *massif* as *Chomolungma* without singling out any single peak. In 1951, in a blaze of nationalistic feeling, the Nepalese government adopted the name *Sagarmatha* for the peak, but this admittedly poetic name has not caught on, except in Nepal. Thus, the colonial name Mt. Everest has stood up above all challenges as serenely as the peak itself stands up over its lofty neighbours.

As a footnote to this story, the height of Mt. Everest was measured again by the Survey of India in 1954 and found to be 29,028 ft. In 2015, there was a massive earthquake in Nepal, after which some Himalayan peaks lost several feet in height. A new measurement of Mt. Everest's height was made by a joint team from Nepal and China, and their result was declared in 2020. It is 29,032 ft. It seems that the world's highest mountain has grown by 30 feet since the days of Radhanath Sickdher.

2.5   'The Man Who Discovered Mt. Everest'

A more bitter and often unseemly controversy rages around the credit for the discovery of the highest mountain on Earth. As the above story shows, the discovery was the result of admirable teamwork by a group of highly competent scientists, led from the front by Andrew Waugh. If any single individual deserved to have the mountain named after himself, it was probably the dogged Surveyor-General. And yet, every so often, the claim arises that the mountain should have been named after the Chief Computor, and should really have been called Mt. Sikdar. In these stories, Radhanath is said to have been a youthful subordinate who burst into his superior's office one day with the astonishing claim "*Sir, I have discovered the highest mountain in the world*!" And then the scheming *sahib* patted the 'native' boy on the back and craftily stole all the credit, naming the peak after himself—or so the story runs, fusing Everest and Waugh into a single evil *sahib*. This has all the elements of a Bollywood film, but it just didn't happen that way. As a matter of fact, Sir Andrew

Scott Waugh was a man of scrupulous honesty who gave Radhanath and each his subordinates their full due when reporting the discovery. If anyone is wronged by these stories, it is the god-fearing Scotsman. Perhaps the most fitting conclusion to this part of the tale is a sentence from a letter by Waugh to Radhanath, which says "*I am glad that the name I have given to the highest snowy peak has given satisfaction to yourself as well as other superior members of the Department.*"

It was not Waugh, but his successor Henry Thuillier—the man who had announced the great discovery to the Asiatic Society—who tried to airbrush Radhanath's name out of the Survey's records. It makes an unedifying story, but it happened years later, after Radhanath's death. In fact, the cudgels for Radhanath were taken up by Bengali nationalists only in the wake of the unpopular 1905 partition of Bengal. It was they who made out of the dead mathematician a legend and a martyr—and more importantly, a stick to beat their British rulers with. Claims, counterclaims and polemics still resound around the issue. This is, unfortunately, the Radhanath who survives in the popular eye as 'the man who discovered Mt. Everest but was denied the credit'.

The real Radhanath, however, continued to work quietly in the Survey of India's Calcutta office till 1862, when he retired, aged 50. His long years working with foreigners had alienated him from his own middle-class Hindu roots, and his conversion to Christianity in 1854 no doubt accelerated the process. There was to be no homecoming to Sikdarpara for him. Radhanath, therefore, chose the French settlement at Chandernagore (Chandannagar) and settled down there at Gondolpara, using his lifetime savings to build himself a comfortable villa on the banks of the River Hooghly. In 1870, he died and was buried in Chandernagore. No one knows exactly where his grave lies.

Radhanath Sickdhar had proved that Indian brains are quite as good as European brains in their own game of modern science. There lies his true legacy, not in a pinnacle of snow-covered rock.

Chapter 3
Reformers and Educators

The phenomenon of Radhanath Sickdhar would not have been possible, had he not received a solid grounding in western-style education from his teachers in the Hindoo College, even apart from Derozio and Tytler. The fact that Radhanath was followed by other Indians, in Bengal and across the country, was only made possible by the growth of 'English' education in British India. This is an interesting and intricate tale, meriting several volumes in itself, but it bears a brief retelling.

3.1 The Mohammedan College of Calcutta

The ascendancy of the British in Bengal started, as we have seen, with the victory of Clive at Plassey in 1757 and the domination of a merchant company over an entire province. In the evocative words of the poet Rabindranath Tagore, *"The trader's measuring rod presently reappeared as a kingly sceptre"*. It has been mentioned that Parliament passed a Regulating Act in 1773. Under this Act, among other things, the Governor of Bengal was renamed Governor-General. The first man who would hold this position was Warren Hastings.

Warren Hastings (1732—1818) was an old India hand, having been in Calcutta as early as 1750. He now found himself tasked with bringing order out of chaos, and good governance out of a system of plunder. He knew that his English officers were unfamiliar with the country and its languages. The older ones were steeped in sybaritism and corruption. Thus, it was to young British youths, fresh from the playing fields of Eton and the austere Kirk of Scotland, that the Governor-General

S. Raychaudhuri, *The Roots and Development of Particle Physics in India*, SpringerBriefs in History of Science and Technology, https://doi.org/10.1007/978-3-030-80306-3_3

looked to create the just State he dreamed of.[1] To implement this, they would have to understand the laws and mores of the people they would be governing.

Bengal had been under the domination of Islamic rulers for six centuries and the language of administration, therefore, was that of a Mohammedan state, with Urdu and Persian being the main secular languages, while Arabic was the language of the Quran and the Islamic laws. The legal apparatus was also basically Islamic. However, the bulk of the population consisted of Hindus, who clung to their ancient faith and social laws. An uneasy truce had been forged between these two groups. Into this finely-balanced society, the callow British youths brought with them the simplicity and naïveté of their native country, and a sense of entitlement, coupled with an overwhelming desire to get rich as fast as possible. Most of them were 'good Christian men', who could empathise with the iconoclastic Mohammedans, but found the ancient Hindu religion quite disturbingly alien, and were rather shocked at the idol-worship which is central to the Hindu worship religion. The intricacies of the caste system they found bewildering and they found ridiculous the dietary taboos of Hindus and Muslims alike. They felt insulted when women were secluded from their gaze and when they learnt that high caste Hindus regarded their very presence as polluting. And these youths made no secret of their feelings.

The situation was indeed explosive, and Hastings knew it. He therefore determined that a new class of 'agents' and minor officials would be trained by the Company itself to provide his 'boys' with the right kind of assistance in performing their proper duties. In 1780, therefore, Warren Hastings set up what he called the Mohammedan College of Calcutta, and what its first head, *Mulla* Majuddin, called the *Aliah Madrasah*, to impart a proper education to Muslim youths who would help in the administration of the nascent British power. The Mohammedan College taught Arabic, Persian and Islamic Law to its students and issued to them diplomas attesting their ability to administer law and work in the revenue service. Later the syllabus expanded to include other subjects—Arithmetic, Philosophy, Logic, Rhetoric, Theology and even a bit of Science. However, the blinkered view of the clerics who ran the College prevented it from really taking off. They continued to hark back to earlier times, and steep the students in Persian and mediaeval literature, while the Industrial Revolution was playing out in Europe. Soon the race for modernity would be taken up by the Hindus, and they would surge ahead in picking up the science of the *firangis*.

The Mohammedan College, though it now fades away from our story, continued to survive in different avatars. After 1857, it was downgraded to an ordinary school with a separate *madrasah* unit teaching only Arabic, Persian and Muslim Law. By then it was known as the 'Calcutta Madrasah' and it remained a training ground for aspiring Muslim youths. After Indian independence, some of its staff moved to Dacca (now Dhaka), in the then East Pakistan, and founded the Government *Madrasah-e-Aliah* of Dacca, which claims to be the true *Aliah Madrasah*. The mother institution continued

---

[1] Sadly, Hastings did not apply these high standards to himself. His impeachment before Parliament cited his extensive corruption and crimes against the people of India. Though he was formally acquitted, Hastings became a social outcast and died in poverty.

to function in Calcutta, however, but its standards declined over the years and by the early days of the twenty-first century, there were urgent calls for government intervention to save India's oldest college. The Government of the state of West Bengal, where Calcutta is situated, did comply, and in 2009, elevated it to the status of a university, renamed the Aliah University, in a curious fusion of the traditional and the modern.

3.2 The Fort William College

Warren Hastings was succeeded by Lord Cornwallis (1738–1805), the soldier whose surrender at Yorktown had sealed the fate of the British colonies in America. Cornwallis proved to be far more effective as an administrator, and will be forever remembered for his 'Permanent Settlement', which created the *zamindari* or landlord system in Bengal and elsewhere. Whatever the merits or demerits of this system, many *zamindars* proved immensely generous in supporting education and science.

Cornwallis was succeeded in 1796 by Richard Colley Wellesley (1760–1842), a much younger man, who, after Clive and Hastings, may be considered the third founder of the British Empire in India. He followed a policy of aggressive expansion—but wars are expensive and they nearly emptied the Company's hitherto well-filled coffers. Wellesley needed to make his revenue administration more efficient and to that end he decided that his (or rather, the Company's) civil servants needed to learn some Indian languages and laws themselves and not rely on intermediaries such as the Mohammedan College-certified agents. To this end, he founded a college for civil servants at Calcutta. Initially known as the Oriental Seminary, it soon became the simpler Fort William College. This was started in 1800 and shut down in 1854. During this half century, it played a seminal role in the development of modern Bengal and indeed, modern India.

The professors at the Fort William College were initially mostly Europeans who had made a study of Indian languages and society but Wellesley also brought in a number of learned Indians. Eventually, the College developed in a startlingly different way from what its founder had intended. The imperious Lord Wellesley had set it up without consulting the Directors back in England, and they, led by the equally imperious Charles Grant, were deeply suspicious of the new creation. The Directors did not, indeed, achieve the closure of the Fort William College—for Wellesley fought back—but they undercut his efforts by founding in England a parallel institution known as the East India Company College, at Haileybury in Hertfordshire. Here boys of age 16–18 years were well drilled in the idea of European supremacy, and what Kipling famously described as '*the white man's burden*', even as they picked up a working knowledge of Indian languages.

A surprisingly large number of the 'Haileybury boys', as they were called, did enrol again in the Fort William College, but nevertheless the focus of the College shifted from teaching languages to writing textbooks, creating grammars and providing vernacular tracts for missionary activity. It was in the fires of Fort

William College that modern Bengali prose was forged, and it was here that Hindi and Urdu were differentiated and formalised into two different languages out of the chaotic 'Hindostani' which was spoken across wide swathes of India. Moreover, it fostered Hindu intellectuals and reformers like Ishwar Chandra Vidyasagar, who spearheaded the Bengal Renaissance. It did take some steps towards becoming Wellesley's 'Oxford of the East' before it was made redundant by Lord Bentinck in 1830–31, when he made English the official language of instruction. Eventually, Lord Dalhousie provided the *coup-de-grace*, shutting down the Fort William College in 1854.

Science was very perfunctorily taught in the Mohammedan College as well as in the Fort William College. The entire stress was on languages and literature. Step by step, however, the Industrial Revolution caught on, and with them came an increased respect for science. The most important role played by the Fort William College was to engage with the Hindu community through their respected pundits and gradually make them interested in acquiring Western-style education. This then led to an efflorescence of science.

3.3   The Orientalist Movement

Before we get to science, however, it must be mentioned that it was not that Indians alone strove to acquire knowledge of European learning. It was, in fact, a two-way traffic, as the British also tried hard to learn about the vast country which was now theirs to govern. For the westerners, there was much that was mysterious, and much that was fascinating in this new country. The nineteenth century, which saw the establishment of several clubs and societies devoted to science, on the model of the Royal Society and the French Academy, also saw the foundation of several such organisations whose sole purpose was to understand the ancient country where the colonials found themselves in government. Perhaps the very first of these in India was Sir William Jones' *The Asiatick Society*, founded at Calcutta in 1784, with the lofty mandate *"The bounds of investigations (sic) will be the geographical limits of Asia, and within these limits its enquiries will be extended to whatever is performed by man or produced by Nature."*

The Asiatic Society (as it was spelt after 1825), moved to its present building in 1808, and is still housed there. Though it is noted more for its work on unravelling Indian history and archaeology, a great deal of the initial work was connected with the geography and natural history of the Indian subcontinent and with investigating ancient Indian achievements in chemistry, metallurgy, astronomy, mathematics and medicine. The story of this great institution requires a volume in itself, but it may not be out of place to briefly mention some of its leading lights.

Sir William Jones (1746–94), the founder of the Society, was a judge with a flair for languages, and a love of scholarship. Within a year, he had founded the Asiatick Society, and started learning Sanskrit. While learning this liturgical language of Hindu India he was struck by its similarities with Latin, Greek and Persian, all of which he happened to know well. Though not the first to postulate a common origin for these, he was definitely its first influential proponent. This class of languages is now known as Indo-European and their common source is now known as Proto-Indo-European. Sadly, this brilliant scholar died at the young age of 47, and lies buried in Calcutta, the city where he performed his prodigious works of scholarship. After Jones, the baton passed to Henry Thomas Colebrooke (1765–1837), a Sanskrit scholar who is best remembered for his translation of the laws of Hindu inheritance enshrined in the *Mitakshara* of Vijnaneswara (fl. 1100 CE) and the *Dayabhaga* of Jimutavahana (fl. c. 1150 CE), which, though flawed by improper understanding of the ancient society, were the first time Westerners tried to penetrate the intricacies of Hindu law. However, by 1817 he had also translated the *Brahmasphuta Siddhanta* of Brahmagupta (628 CE), which was the first introduction of Western scholars to the great advances in mathematics and astronomy made in ancient India. These were works of solid and meticulous scholarship, but were overshadowed by the brilliant achievements of James Prinsep (1799–1840), whose decipherment of the *Brahmi* and *Kharoshthi* scripts form the stuff which legends are made of. Prinsep burnt his candle at both ends, however, and died exhausted at the early age of 41. It was left to Horace Hayman Wilson (1786–1860) to translate the Vedas and the Puranas, the ancient scriptures and quasi historical records of the Hindus. It was also under Wilson's leadership that the Asiatic Society opened its doors (1829) to Indians. Dwaraka Nath Tagore, Sib Chandra Das, Baidyanath Roy, Bunwari Govind Roy, Kalikrishna Bahadur, Raj Chunder Das, Ram Comul Sen and Prasanna Coomar Tagore—all wealthy landowners and cultured dilettantes, were among the favoured few.

Inspired by the Calcutta society, the Asiatic Society of Bombay (now Mumbai) was founded in 1804 as the Literary Society of Bombay and later merged with the Bombay Geographical Society and the Bombay Horticultural Society. Its founder was Sir James Mackintosh (1765–1832), then Recorder of the Bombay Presidency. In 1829, this society, with the sister societies in Calcutta and Madras, were invited to become chapters of the Royal Asiatic Society of Great Britain. In 1832, the Asiatic Society of Bombay moved to the magnificent building which it still occupies today, and whose iconic steps are a favourite place for shooting the 'Bollywood' movies which today's Mumbai is noted for.

The Bombay Natural History Society was set up in 1883. In Calcutta, the Agricultural and Horticultural Society of India was founded in 1820. Similar societies bloomed in all the three Presidencies. Of particular note are the Madras Literary and Scientific Society (1805) and the Calcutta Medical and Physical Society (1823). A few copycat associations were even set up in some princely states which had English-educated rulers, though it is only fair to mention that these short-lived societies were mostly concerned with teaching 'natives' the correct way to hold a knife and fork and how to use toilet paper instead of water.

Apart from discovering and preserving much of the history and archaeology of India, the scholarship promoted by these societies set off the nineteenth century artistic and literary movement in Europe and America which is called *orientalism*, a school of thought which encompassed the scholarship of Schlegel and Max Müller, the romanticism of Schopenhauer and Byron, the sensualism of Delacroix and Gérôme, the pseudo-benevolence of Kipling, the muddled mysticism of Blavatsky and Olcott and the opulent melodies of Holst, Debussy and Rachamaninoff. This movement has, however, been panned by twentieth-century thinkers like Said, Gramsci and Foucault as a western caricature far removed from the reality. Not surprisingly, post-colonial historians find much of it to be offensive or even down-right insulting—which is perhaps an over-reaction. Orientalism is not central to this volume, but in all honesty it must be said that Western interest in India would have been scant if the ground had not been prepared by the orientalists, with all their misplaced romanticism.

3.4   The Hindu *Mahapathshala*

Rammohan Bannerji (or, in the style introduced by the pundits of Fort William College, Bandyopadhyay) was a Bengali Brahmin of high caste, scion of a wealthy family in the service of the Bengal Nawabs for three generations, which entitled then to the honorific *Ray-rayan*. Born in 1772, the young Rammohan was sent first to a Hindu *pathshala* to learn Bengali and Sanskrit, then to a *madrasah* at Patna to learn Arabic and Persian, and finally to Benaras, the heart of Hindu scholarship, to become proficient in the Hindu scriptures. Introduced to the Quran and the works of Sufi philosophers during his *madrasah* days, the teenaged boy was already inclined to monotheism and expressing his opposition to idol worship.

Grown into manhood, Rammohan found employment with the East India Company under the name Ram Mohun Roy, and made a fortune by lending money to reckless Company official who were living beyond their means. At 42, he could retire and settle down in Calcutta, where he soon founded the *Atmiya Sabha* or Friendly Society (1815). This was not a formal organisation, but a private socio-philosophical discussion society. It attracted cultured *zamindar*s as well as Baidya Nath Mukherji, a close friend of Ram Mohun's, for they had worked together for the East India Company. The *Atmiya Sabha* also attracted some mavericks, such as the Scottish watchmaker David Hare, who had come to India in 1800 and made a great deal of money selling watches, but at the same time had been deeply moved by the plight of common Indians and wanted to do something lasting for their improvement.

One day, in April 1816, Hare was present at a session of the *Atmiya Sabha*, where the conversation turned, as it almost always did, on how to rid Hindu society of its backwardness and superstition, without actually embracing Christianity *en masse*. Ram Mohun Roy himself was in favour of founding a reform movement in Hinduism, but David Hare argued that the need of the hour was for Western education and the ideals of the European enlightenment to reach India. He therefore proposed setting up a school and a college where Western education would be imparted without the trappings of the Christian faith. This found general favour. David Hare and '*Baboo Buddhi Nath Mookerjya*' (sic) were entrusted with writing a white paper on this and taking it to a sympathetic person in the upper echelons of power.

The sympathetic person chosen by Baidyanath Mukherji was the Chief Justice of the Calcutta High Court, Sir Edward Hyde East (1764–1847). He had come to Calcutta to take up this important position in 1813 and was well acquainted with the provisions of the East India Company Act of 1813, better known as the Charter Act. The Act contained a clause that a sum of one lakh of rupees should be spent over the next 20 years on "*the introduction and promotion of a knowledge of the sciences among the inhabitants of the British territories in India.*" This was a complete break with the earlier policies of the Company, which had hitherto shown no interest in education, and marked the beginning of a greater involvement of the British Government with their conquests in India. Though nothing much had been done since the passing of the Act, it meant that when *Babu* Baidyanath approached the learned judge with Hare's paper, his ideas fell on fertile ground.

The judge *sahib*, in fact, took up the cudgels with energy and authority—with the full support of the Governor-General, Lord Moira. On May 14, 1816, a meeting occurred at the house of the Chief Justice. Among the fifty Indian attendees were *Babu* Baidyanath as well as *Raja* Radha Kanta Deb of Sovabazar, doyen of the ultra-conservative section of the Hindu *bhadralok*. Prominently absent were Ram Mohun Roy and David Hare. In fact, the attendees collectively declared that no subscription should be accepted from 'Ramohin Roy', the infidel. Despite being marred by such personal animosity, however, the resolution to found a college was enthusiastically passed and nearly 50,000 rupees were raised on the spot, more being promised. In fact, another 12,000 rupees would soon be donated by the *Maharaja* of Burdwan—the biggest donor of them all. The largely orthodox Hindu Management Committee consisted of two Governors—*Raja* Pratap Chandra of Burdwan and Gopi Mohan Tagore of Pathuriaghata, and five Directors—Ganga Narayan Das, Gopi Mohan Deb (father of Radha Kanta Deb), Radha Madhab Banerjee, Jay Krishna Singh and Hari Mohan Tagore. There was a European Secretary, Lieutenant Francis Irvine and the Native Secretary was, of course, the ever faithful *Babu* Baidyanath.

On a cold, wintry Monday morning on January 20, 1817, classes started at the Hindoo (*sic*) College, with 20 students, all from Hindu *bhadralok* families, receiving instruction at rented premises in the Garanhata area of north Calcutta. At this stage there were two sections—the junior (*pathshala*) and the senior (*mahapathshala*)—and each had only language classes in English, Bengali, Persian and Sanskrit. The teachers were mostly Europeans and the educational system closely followed the style of the missionary schools and the Fort William College, except that there was no religious or moral instruction. Education at this elite institution was expensive—5 rupees per month, equal to the full monthly salary of a schoolteacher. Nevertheless, the number of students had grown by 1819 to 70, by 1826 to 196 and by 1830 to 409. The College premises moved around to three locations in the first eight years, until David Hare donated a piece of land just north of the Gol Dighi (now College Square) where a school building was built in a beautiful neo-classical style. The Hindu School (which developed out of the *pathshala*) still stands there, though the old building has been replaced by a high-rise modern monstrosity. In 1855, the *Mahapathshala* was separated out and moved to a new building across the adjacent road (now College Street), and this is where it stands now.

In 1824, seven years after its foundation, the College, where David Hare was now a 'Visitor', managed to find a science teacher in David Ross, an engineer from the Calcutta Mint, who taught an eclectic combination of many subjects, including Astronomy, Geography, Chemistry, Mineralogy and Optics. Ross, however, was not much of a success as a teacher, earning among the students the nickname 'soda *sahib*', referring to his frequent references to the word 'soda'. But when the College, in 1828, employed Dr. John Tytler as Professor of Mathematics, his lectures managed to fascinate the students and earn him a wide fan following. Notably, it was this same Tytler who recommended Radhanath Sickdhar and his friends to the Survey of India. In the same year came Derozio, and the College became the hub of the Enlightenment in India.

3.5 Bentinck and Macaulay

Around the time when Tytler and Derozio were joining the *Mahapathshala*, the East India Company was going through a deep financial crisis. By the 1820s, much of India had come under its sway, with victorious campaigns being fought all over the subcontinent. However, the cost of these wars was ruinous, and by 1825, the East India Company found itself deeply in the red. In 1825, the Company's annual losses amounted to 1.5 *million* pounds. Not surprisingly, the Company's shareholders—many of whom were MPs—were breathing fire and brimstone down the necks of the Directors. In desperation, the harried Directors turned to a different stamp of man. This was Lord William Bentinck. An aristocrat to the fingertips, he was a maverick, an interventionist and a man who put his conscience above the orders of his superiors. Thrice he had held high commands—including Governor at Madras—and thrice he had been recalled for following his own policies. When his name was proposed as the man who would reform India and bring the East India Company back to solvency, the Prime Minister may have been only too happy to pack him off as far as possible from London. It was a momentous decision, for Bentinck was to change India—in many ways, forever.

The new Governor-General landed at Calcutta in 1827—a man with a mission, or rather, two missions. The one which concerns us was to turn around the finances of the East India Company, and this indeed he did, for by the time he left, the Company was registering a *profit* of 1.5 million pounds a year. This was achieved by a slew of measures, including strict economies and policies to increase the revenues. Bentinck quickly realised that mere collection of agricultural taxes and occasional spoils of war could not sustain the Company in its basic programme of extracting wealth from India. It was necessary, reasoned the astute statesman, to extract the *natural* wealth of India from its sources, using the tools available to modern science and industry. And thus, in 1830, he revived the Survey of India, recalling George Everest to take the helm, as we have described earlier. Just before leaving India in 1835, Bentinck formed the 'Coal Committee' to investigate the coal reserves of India, for the Age of Steam had begun. This grew into the Geological Survey of India (1851). His patronage of Dr. Nathaniel Wallich at the Botanical Gardens outside Calcutta may be regarded as the beginnings of the Botanical Survey of India. The same savant was also encouraged to augment the collection of stuffed animals at the Indian Museum in Calcutta, an activity which later grew into the Zoological Survey of India. However, the reform made by Bentinck which had the most far-reaching effects came only in 1835. This was the replacement of Persian by English as the official language of instruction in Government institutions—the real watershed between Mughal India and British India.

That Indians should be instructed in the learning which *he* believed to have most value was always there in Bentinck's mind, but his initial years were a struggle to improve the Company's finances and overcome orthodox opposition to his social reforms—the second string to his India mission. Educational reforms, therefore, went on the back burner, and the whole machinery of government continued to creak along on the twin wheels of Arabic and Persian. Moreover, his Council did not like his interventionist mindset, but preferred a *laisséz faire* policy of 'masterly inaction'. In the last year of his term, however, there appeared on the scene a vigorous new personality who was even more passionate about 'educating the natives' than the Governor-General himself. This was Thomas Babington Macaulay.

Born in 1800, Macaulay was a child prodigy with a particular felicity for languages. By the time he graduated from Trinity College, Cambridge, he could write poetry with equal felicity in English, Latin and Greek. He was also supremely eloquent in English prose and has left us with ringing phrases which resound down the corridors of history. Such manifest skills inevitably led him to politics. Appointed as the Law Member on the Governor-General's Council in Calcutta, this eloquent, passionate and arrogant scholar of 34 years landed at Calcutta in February 1834, to join his reforming ardour with that of his reforming superior, the Governor-General.

Among the British intellectuals in India, there were two factions. The 'Classicists' had a partiality for Indian classical learning, while the 'Modernisers', felt that financing oriental studies was a *'bloody waste of money'*. Their dispute reached its peak in the time of Bentinck, who was primarily a soldier and not a scholar. Unable to tolerate the incessant squabbling of these erudite ones, he made the newly-arrived Macaulay his right-hand man, leaving these decisions to his superbly self-confident aide. For indeed, Macaulay had no doubts at all—about this, or about anything else in his life. The urbane and worldly-wise Lord Melbourne, soon to become Prime Minister, once remarked with typically dry humour *"I wish I was as cocksure of anything as Tom Macaulay is of everything"*.

Indeed, with Macaulay, there was never any question as to which way the dice would fall. In a powerful speech in the House of Commons supporting the Charter Act of 1833, he had already taken the moral high road, declaiming *"Are we to keep the people of India ignorant in order that we may keep them submissive? Or do we think that we can give them knowledge without awakening ambition? Or do we mean to awaken ambition and to provide it with no legitimate vent? Who will answer any of these questions in the affirmative? Yet one of them must be answered in the affirmative, by every person who maintains that we ought permanently to exclude the natives from high office. I have no fears. The path of duty is plain before us…"*

He went even further to say *"It may be that the public mind of India may expand under our system till it has outgrown that system;… that, having become instructed in European knowledge, they may, in some future age, demand European institutions. Whether such a day will ever come I know not. But never will I attempt to avert or to retard it… The sceptre may pass away from us… Victory may be inconstant to our arms. But there are triumphs which are followed by no reverses."*

His words were to prove prophetic,[2] for 114 years after he had delivered his speech, the British were to withdraw from India, leaving the land to be governed by 'natives' trained in the Western style, many of whom had actually finished their education in England.

This, then, was the idealist who was now tasked with determining the future course of education in British India. Macaulay threw himself with gusto into his new role, producing, in February of 1835, his famous '*Minute on Indian education*' in which he argued with all the force of his eloquence against the orientalists in the Committee. Discussing the arguments between the two factions, the 'Minute' is actually a polemic masquerading as a judgement. With typically insular arrogance, Macaulay declared *"It seems to be admitted on all sides, that the intellectual improvement of those classes of the people who have the means of pursuing higher studies can at present be affected only by means of some language not vernacular amongst them… We must teach them some foreign language. … Whether we look at the intrinsic value of our literature, or at the particular situation of this country, we shall see the strongest reason to think that, of all foreign tongues, the English tongue is that which would be the most useful to our native subjects."* When this Minute reached Lord Bentinck, he found the views of the Scottish idealist in total consonance with his own. And so he passed the English Education Act of 1835, which said *"His Lordship in Council directs that all the funds… at the disposal of the Committee be henceforth employed in imparting to the native population a knowledge of English literature and science through the medium of the English language."*

And that was that. The British Government in India would not, it is true, close down the colleges of oriental learning it had already established, but in future all money was to be spent only in teaching the English language and 'English' sciences. After Bentinck, his successor Lord Auckland, another committed Whig, passed Act XXIX of 1837, which made English the official language of the law courts and all Government business. This Act did more to implant English firmly among the Indian population than all the idealism of Bentinck and Macaulay, for it made English the road to a well-paid Government job. For, after all, 'where the purse goes, the heart is bound to follow'.

Macaulay, and others of his class and time, genuinely (if not wholly accurately) thought of the upper class Englishman as the highest form of humanity, and their wish to create 'brown Englishmen' was not malicious, but kindly and

---

[2] For those modern historians who see in Macaulay's proposals for English education in India a diabolical plan for the permanent enslavement of Indians, this speech should come as an eye-opener. As a historian, Macaulay was particularly well acquainted with the impermanence of dominion. There is no reason to think that his opinions were not sincere.

idealistic—if unabashedly patronising. As a matter of fact, this has largely been the educational policy of India ever since, even after seven decades of Independence.

As for Macaulay himself, he returned to England in 1838, rich and famous, for the story of his 'Minute' had spread through all England. Created a peer in 1857 as Lord Macaulay of Rothley, he died two years later and was buried with great pomp and ceremony in Westminster Abbey. Lord Bentinck had preceded him to the grave by two decades. But their work lives on.

3.6   De Morgan's Ramanujan

After Radhanath Sickdher, the flame of science in India burned low, especially in Calcutta, which produced many erudite youths, but no real spark of originality till the closing years of the century. Across India, as western learning caught on, men (and a few women) began to master the science of the westerners and teach it in the burgeoning number of colleges, which we shall describe in the next section. In the midst of this appeared a highly-talented mathematician, whose life was as chequered as it is forgotten today. Indeed, if anyone may be said to have carried the spirit of Radhanath Sickdher forward, it was this man. He was known to the British as 'Master Ram Chundra'.

Ram Chundra Lall or Ramchandra Lal Mathur, to give him his proper name, was born in 1821 at Panipat, near Delhi. The son of a revenue clerk in the service of the East India Company, he was one of the first fruits of the new education policy of the Charter Act of 1813. He was educated at the Delhi English Institution, which had been set up by Sir Charles Trevelyan, chief assistant to the British Resident at the vestigial Mughal court, in 1828. This pioneering institution provided the first introduction to English, as well as Western mathematics and science, to scholars from North India. It still survives as the Zakir Husain College (now proudly renamed the Zakir Husain Delhi College). In the initial days, just as the Hindu *Pathshala* and *Mahapathshala* ran as sister institutions, so did the Delhi English Institution and the Delhi College. Thus it was that, after graduating from them in succession, the bright young Ram Chundra, after spending two years as a junior clerk, was appointed as a junior teacher of 'European science' in the Delhi College (1844).

We do not know very much about 'Master Ram Chundra' as a teacher, for he seemed to have been more a Derozio than a Tytler. A major programme at the Delhi College was the 'Vernacular Translation Society' whose mandate was translation of Western (mostly English) books of science into languages then widely in use by the educated Indians, namely Arabic, Persian and Urdu. Ram Chundra threw himself into this work wholeheartedly. But he did not stop at mathematics, emulating Derozio in promoting Western ideas and debunking traditional beliefs. In his own words, *"I... formed a society for the diffusion of knowledge among our countrymen... We first commenced a monthly, and then a bi-monthly periodical... in which notices of English science were given, and in which not only were the dogmas of the Mohamedan* (sic) *and Hindu philosophy exposed, but also many of the Hindu superstitions and idolatries were openly attacked. The result of this was that many of our countrymen, the Hindus, condemned us as infidels and irreligious; ...."*

In 1850, Ram Chundra took advantage of a three-month 'examination leave' at his College to travel to Calcutta and publish a book on mathematics at his own expense. Calcutta was then the only place where he could find a press with the equipment to print mathematical symbols. Ram Chundra's book, which was in English, was called "*A Treatise on Problems of Maxima and Minima, solved by Algebra*". Its purpose was to find algebraic methods for finding maximum and minimum values in problems of number theory and coordinate geometry, without the use of the techniques of differential calculus, which are (then and now) universally taught in schools and colleges.

One can get a flavour of Ram Chundra's style by looking at the very first problem taken up by him, viz. '*To divide a given number into two such parts that their product may be the greatest possible*'. In modern language, the problem is to take a given (real) number a and partition it into two (real) numbers x and $a - x$ such that the product $p = x(a - x)$ is maximum. The calculus method of solving this is well-known: we set $dp/dx = 0$, which yields, on differentiation, $a - 2x = 0$. We then solve this to get $x = a/2$. Instead of doing this, however, Ram Chundra defines $x = a/2 + y$, and substitutes this in $p = x(a - x)$ to get $p = a^2/4 - y^2$, which is clearly maximum when $y = 0$, i.e. $x = a/2$. This is rather clever even if it is trivial. However, this is just the first problem—a sort of warming up exercise to his method. The problems grow tougher as the book progresses, and the entire work is filled with similar sleight-of-hand techniques, making one marvel at the ingenuity of the author. He modestly claims in the preface that such methods might be useful to students '*who are not advanced in their study of the Differential Calculus*'.

It is not surprising that Ram Chundra's work failed to draw immediate plaudits. The Calcutta Review (July–December 1850) panned his efforts,[3] writing "*It is with sincere regret that we are compelled to speak with very limited approval of the merits of this work,... While we are thus compelled to express our doubts, as to the utility of the object of the book, we cannot be much more complimentary as to the mode of its execution, which is, in general, clumsy and school-boy-like.*" Despite the ungracious, even churlish, tone of the review, the criticism on 'usefulness' is not quite baseless, for even the simple example given above will show that Ram Chundra's methods cannot be easily duplicated by a less nimble mind. This, however, is analogous to complaining about a beautiful painting because it cannot be reproduced by a house painter.

Stung by this diatribe, Ram Chundra tried to defend himself in '*The Englishman*' (later re-named '*The Statesman*'), but the damage had been done, and even today his work remains an almost-forgotten curiosity. However, the controversy did attract the attention of a prominent member of the Governor-General's Council, namely John E. Drinkwater-Bethune, a friend of Indians, who will be introduced in the

---

[3] The full review may be found at the link http://www-groups.dcs.stnd.ac.uk/history/Extras/Cal cutta_Review_1850.html.

next section as the founder of Bethune College.[4] This kindly gentleman bought
up 37 copies of Ram Chundra's book (for 200 rupees), and undertook to send it
to various mathematicians and scholars in England for their opinion. We do not
know the fate of 36 of the copies, but one of them reached the famous logician
Augustus de Morgan who, having been born in Madurai, had a better impression
of the Indian mind than most of his compatriots. De Morgan, then Professor of
Mathematics at the University College, London, was thrilled to read this work, where,
for all its *naiveté*, he recognised genuine mathematical talent. In 1859, he ensured
that the book was re-published in England, with a preface which he wrote personally,
including autobiographical notes received from Ram Chundra himself. Recognition
by De Morgan made the unassuming science teacher from Delhi shoot up in people's
estimation, at least among the British in India, and thus Ram Chundra has been
described as 'De Morgan's Ramanujan'.

In May 1852, Ram Chundra was received into the Anglican Church. Presumably
this brought him peace of mind during the years while De Morgan was working
through his book. Exactly five years later, however, in May 1857, the Great Rebellion
broke out and soon Delhi fell under the control of mutinous sepoys, who ran riot
for several months. Inside Delhi, Indian converts to Christianity were viewed as
collaborators of the hated British, and were murdered in cold blood by the raging
soldiers. Ram Chundra himself survived and made his escape from Delhi to the British
camp outside the walled city after some hair-raising adventures. He was immediately
employed as an interpreter to the British army, and as such, accompanied them on
their recovery of Delhi in September 1857—and their subsequent brutal suppression
of the revolt. He was rewarded with a couple of headmasterships in quick succession,
once peace was restored, but in 1866, Ram Chundra retired from the Company service
and moved to the Patiala State, where he was immediately employed by the forward-
looking *Maharaja* Mahendra Singh, who made him Director of Public Education
of the Patiala State (1870). He continued to work tirelessly for the furtherance of
education, and also that of his adopted religion, till continuing ill-health brought
about his demise in 1880, aged fifty-nine.

To come back to mathematics. De Morgan, of course, knew that Ram Chundra's
work was not of the kind which would take the frontiers of mathematics forward, but
instead he realised that the Indian's techniques were peculiarly suited to induce
nimbleness of mind in young mathematicians. As such, he was active in introducing many of Ram Chundra's problems into the curriculum of English schools
and colleges. Many of these entered into the famous Victorian mathematics textbooks by Isaac Todhunter, a student of De Morgan, and thence they have percolated
into school texts around the world. Thus, generations of schoolchildren have strug-
gled to solve some of the notoriously difficult problems in Todhunter's '*Algebra for
Beginners*', without knowing that they were actually being asked to rediscover some
of Ram Chundra's clever manipulations.

---

[4] In a repeat of the Everest story, the founder's surname is correctly pronounced 'beaton' (beat'n),
but in Calcutta the college name is universally pronounced 'bay-thoon'.

Yet 'Master Ram Chundra' is a forgotten figure today. The nationalist movement largely ignored him, probably because he was regarded as a British stooge, though in fact, he was as critical of Christian bigotry as he was of Hindu and Muslim narrow-mindedness. Today, more than a century and a half later, we can safely put by the political upheavals of his time, and give due honour to the first Indian since mediaeval times to make original contributions in pure mathematics.

3.7   Colleges Proliferate Across India

Both Radhanath Sickdher and Master Ram Chundra were, as we have seen, products of the new learning which the British *raj* brought to India. In fact, both of them absorbed more than the science of the foreigners, for they also embraced their religion. Christianity[5] was, in fact, part of the package as these pioneers saw it, and it seemed reasonable to them to go the whole hog and absorb all of it. One can say the same of other leading lights like the poet Michael Madhusudan Dutt, another Hindu College product, and, a little later, *Pandita* Ramabai of Poona. Ram Mohun Roy, on the other hand, with his eclectic interests, managed to dance a tightrope between the pantheism of his fathers and the monotheism of his masters, which he contrived to maintain all his life. Ironically, the syncretic *Brahmo Samaj* which he founded did more to stem the conversion of thinking Indians to Christianity than all the fulminations of his conservative detractors. The *Brahmo Samaj* and its clones, the *Arya Samaj* and the *Prarthana Samaj,* provided a niche where a section of open-minded Indians could stay within the broad pale of Hindu society, while at the same time repudiating some of the customs and rituals which seemed repugnant to their western-trained minds. With the proliferation of colleges and universities around India following the initiative set in motion by the Charter Acts however, Hindu—and somewhat later Muslim—society rose to the challenge and created a *via media* whereby a scholar could absorb the science and learning of the West while retaining the faith in which he or she had been raised. It is important to realise this aspect of the new learning, for European science could never have spread widely in India if it had run contrary to the religious faith of the majority.

As we have seen, the Anglo-Mohamedan College (1780), the Fort William College (1800) and the Hindu College (1818)—all in Calcutta—were the first major colleges to be established in India, followed by the Anglo-Arabic Delhi College (1828).

---

[5] Christianity, it may be mentioned, was not new at all to India. The Apostle St. Thomas is believed to have preached in India and been martyred within the confines of today's city of Chennai (72 CE). Large numbers of Nestorians fled to India during the persecutions that followed the Council of Ephesus (431 CE) and still survive as the 'Syrian Christian' community. The Portuguese, spearheaded by St. Francis Xavier, made large numbers of converts to Catholicism in the small enclaves controlled by them, such as Goa. However, Christianity remained peripheral to the mass of the Indian population, until the vigorous and partly state-sponsored missionary efforts of the Protestants. Even with that, Christianity as a religion has never really caught on in India, except in the north-east and in pockets in southern India.

However, rather than follow a strict chronological order, we shall describe the foundation of new colleges and universities in a zone-wise manner, i.e. in the three Presidencies of Calcutta, Madras and Bombay, representing east, south and west, respectively, and then, separately, for northern and central India.

Let us take up the story from the east, for that has already figured largely in our story. Within a year of the foundation of the Hindu College in Calcutta, i.e. in 1818, William Carey, with his colleagues Joshua and Hannah Marshman and William Ward, started their own college at Serampore (Srirampur)—officially Frederiksnagore—the Danish colony in Bengal. The core purpose of the College was, indeed, the teaching of Christian theology and Christian religion, but as that alone would have secured very few students, it also offered courses in other subjects. In 1827, Joshua Marshman travelled to Denmark and secured from King Frederik VI a Royal Charter giving the Serampore College the status of a University, i.e. enabling it to confer degrees in all faculties. In 1857, after the formation of Calcutta University, the arts and sciences faculties of the Serampore College were re-affiliated to the new University, but the right to grant degrees in theology were retained and confirmed by the Serampore College Act of 1918. In fact, this continues even today.

Lord Bentinck founded the Calcutta Medical College in 1835. Dissection was a major issue, for Muslims are forbidden to desecrate a corpse and the mere touch of a dead body is considered to create ritual pollution in upper caste Hindus. Initially, therefore, the Medical College had to make do with wax models to teach anatomy and the basics of surgery—but that was far from the real thing. Now, however, the taboo was broken by a middle-aged *Ayurveda*[6] practitioner, who had been transferred from the Sanskrit College to the new Medical College. His name was Madhusudan Gupta. In 1836, in a shed within the College premises, Gupta carried out the first dissection by a modern Indian. There was no looking back, though Gupta then had to face a sort of Inquisition from an assembly of Hindu pundits. This he passed with flying colours, ably quoting Sanskrit scripture and ancient texts to justify his act. So successful was his defence that the taboo gradually faded away and medical teaching in India gradually became identical with European styles and standards. Of course, by then, the superiority of European medicine over Indian traditional medicine was becoming apparent to the general public. It was easier for both Hindu and Muslim society to open their doors for treatment to fellow Hindus and fellow Muslims, rather than to an alien from Europe, however skilled. In fact, never was the elasticity and resilience of Indian society so powerfully on display as in this ready adoption of modern medicine.

Perhaps the experience with Madhusudan Gupta inspired the people of Krishnanagar, where the famous debate was held, to embrace Western-style education. In 1845, in response to the demands of his subjects, the *Maharaja* of Nabadwip (Nadia), Srisha Chandra Roy, petitioned the Governor-General, Lord Hardinge, to found a College at his capital, Krishnanagar, and received a favourable response. On January 1, 1846, the Krishnanagar Government College was started with a generous grant

---

[6] *Ayurveda*, or the Veda of Longevity, is a system of traditional medicine which developed in India from ancient times.

from the Bengal Government. The grand neoclassical buildings of the College were ten years in the building and were finally inaugurated in 1856.

Slightly earlier than the Krishnanagar Government College was the Midnapore College, originally founded as a school in 1834. The school still exists, but, like the Hindu School and College, it budded into a new College in 1873. Initially started by rich landlords after the manner of the Hindu College, it was taken over by the Government within two years, i.e. in 1836. In 1840, it became a 'Zilla School', or District School, which entitled it to full Government funding.

In the Calcutta of the 1840s, there was a theatre hall called the Sans Souci Theatre, which stood on Park Street. After a devastating fire, its owner sold out to a group of Jesuit priests, and it was there, in 1860, that Fr. Henri Depelchin, the Superior of the Bengal Mission of the order, decided to found a new college. It was called the College of St. Francis Xavier, and Fr. Jean Devos was its first Rector. After Fr. Devos's return to Belgium, Fr. Depelchin himself took over the Rectorship, and guided the new college from 1864 to 1871. At some stage, the name of the college was shortened to St. Xavier's College, and it is by this name that it survives into the twenty-first century as one of modern Calcutta's leading educational institutions. In 1865, Fr. Depelchin recruited a young priest named Eugène Lafont to teach science at the fledgling college. A brilliant teacher, Fr. Lafont was to popularise Western science in Calcutta as few others have ever done.

However, the Catholic fathers had long been upstaged by their bitter rivals, the Presbyterian Church of Scotland. It was a Scottish missionary, the Rev. Alexander Duff, who had founded the first missionary-run college in Calcutta proper in 1830. It was initially known as the General Assembly's Institution, and later renamed the Scottish Church College. A great deal of the College's academic programmes were designed on the advice of Ram Mohun Roy. After 1953, the College has been controlled by the Church of North India, the successor in India to the Anglican and Scottish Churches, but the name of the College has not been changed.

During the nineteenth century and the early years of the twentieth, the Bengal Presidency comprised the modern Indian states of West Bengal (where Calcutta lies), Bihar, Jharkhand, Odisha, Assam, Meghalaya and Tripura as well as the neighbouring country now called Bangladesh. Western education began at the centre of power in Calcutta, and as we have seen, began to spread out to neighbouring areas such as Nadia and Midnapore. It also spread further afield. At Cuttack (Kataka) in modern Odisha, a new college came up in 1868, thanks to the efforts of the local commissioner, Thomas E. Ravenshaw, and a generous grant of money from the *Maharaja* of the Mayurbhanj State, Krishna Chandra Bhanjdeo. Initially called the Cuttack College, it was renamed the Ravenshaw College in 1868, in honour of its founding spirit.

Higher education was slower to come to the north-east of the country, but it began with the foundation of the Cotton College in 1901 at Gauhati (now Guwahati) in the modern Indian state of Assam. The moving spirit behind the foundation of this college was Manick Chandra Baruah, a wealthy tea merchant and an active politician, who founded the Assam Association in 1903. Manick Chandra wrote a strong letter (1899) to the Commissioner for Assam, Sir Henry Stedman Cotton, pointing out that *"Assam is the only province which does not have any college."* Cotton was sympathetic, and so was the Viceroy Lord Northbrook, as a result of which the sanction for a 'Gauhati Government College' came within a year. Classes started at the new college—renamed Cotton College by public acclamation—in May 1901, after an inauguration by Cotton himself.

In Calcutta, a few enlightened souls continued to push for women's education, and this was supported by the British administration. In 1849, John Drinkwater Bethune—the kindly benefactor of Master Ram Chundra, who was also Education Member in the Governor-General's Council—was able to open a 'Native Female School' in Calcutta. In 1856, it was taken over by the Government and renamed the Bethune School, in memory of its founder. In 1879, the Bethune College was founded and thus women's education was extended to the graduation level—which was as high as education went in India at the time. The College was—and remains—affiliated to Calcutta University. Its contribution to the education of women in Bengal has been immeasurable.

We now move to the southern parts of India, where the Madras Presidency sprawled over 24 districts with a population of around 45 millions, which was more than the 40 millions populating the British Isles at the time. Though the focus of history falls more on Bengal, which was the centre of British power in the nineteenth century, the very first college in India actually came up in the Madras Presidency. This was the Church Missionary Society College, or CMS College, as it is better known, at Kottayam, in the then-princely state of Travancore, now in the modern Indian state of Kerala. Kottayam is an ancient town, not far from the popular tourist destination Alleppey (Allapuzha) and the Church Missionary Society was an Anglican-Lutheran organisation. The first batch of missionaries came to India in 1814. The CMS School, now College, at Kottayam started in 1815, and therefore, pre-dated the Hindu *Pathshala* by two years. Initially, it taught mathematics, but not science, but then acquired a full-fledged curriculum after it was affiliated to the newly-formed Madras University in 1857.

As in Bengal, the East India Company was late in taking up education, but the French, in their little enclave at Pondicherry (Puducherry) started a medical school named, rather unimaginatively, as the *Ecole de Médicine de Pondichéry*, in 1823. It was the earliest institution in India to take up the treatment of tropical diseases in real earnest, and was initially staffed by doctors from the French Navy and from their worldwide colonial service. The students were given a sort of diploma called *Médicin Locale,* which permitted them to practise within the French colonial empire. In 1956, India and France signed a treaty, under which Pondicherry was ceded to India. With the takeover, the medical school was renamed the Dhanvantri Medical College, Pondicherry. The college had already started postgraduate medical studies. In 1964, it was renamed the Jawaharlal Nehru Institute of Postgraduate Medical Education and Research, which is usually abbreviated as JIPMER.

As in Bengal, Jesuit missionaries were at the forefront in setting up educational institutions. In 1844, they set up St. Joseph's College at Trichinopoly (Tiruchirappalli), or Trichy for short, in the heart of the modern state of Tamilnadu. This college produced an outstanding scientist in G.N. Ramachandran, as well as India's 'missile man', the late President A.P.J. Abdul Kalam. However, its fame has been overshadowed by the college founded in Madras (Chennai) by Fr. François Bertrand in 1925. This is the Loyola College, which ranks today among the top colleges in India and has a long list of prominent alumni.

In parallel with Bengal, the efforts of the Jesuit fathers had been anticipated by Scottish missionaries. As early as 1835, Rev. George Laurie and Rev. Matthew Bowie had set up St. Andrew's School in Madras. In 1837, the Rev. John Anderson, who had been sent out for this purpose, set up the General Assembly's Institution, in parallel with Duff's college in Calcutta, and the St. Andrew's School was merged with the new institution. But the real architect of the College was the Rev. Dr. William Miller, who arrived in 1862 and effectively built the fledgling institution into a great temple of learning. At a later stage it was renamed the Madras Christian College, and under that name it is one of modern India's leading educational institutions.

Tellicherry (Thalassery) is a small port to the north of Calicut (Kozhikode), where in the early 1800s, a British ship was wrecked on the huge rocks off the shore, and from the wreck, a young Englishman of unknown antecedents swam ashore, half-drowned and barely alive. He was cared for by the people of Tellicherry, and decided to make this town his home. The sailor's name was Edward Brennen. Securing a job at the Tellicherry port, this lifelong bachelor devoted his life to good works and the upliftment of the people who had saved his life. When he died in 1859 at the age of 75, it was found that he had left his life's savings to found a free school for the people of Tellicherry. Three years later, Brennen's School was founded and it soon grew into Brennen's College. It still exists today, as the Government Brennen College.

Madras also has its own Presidency College, which was created by public subscription without any missionary initiative. The process was kicked off by Sir Thomas Munro—another Scotsman—a former soldier who had been appointed Governor of Madras in 1819. In 1826, a year before he died of cholera, Munro formed a Committee of Public Instruction at Madras. However, with the death of Munro, and the momentous decision made at Calcutta in 1830 that Government funds would be reserved only for English instruction, the Committee of Public Instruction was replaced by a 'Committee for Native Education'—the small change in nomenclature signifying the increasingly paternalistic approach of the colonial government in India. In the following year, the Government posted as Governor of Madras a Scottish nobleman, Lord John Elphinstone. It was also at this time that the Indian Education Act passed by Lord Bentinck was sent to Madras. The new Governor, young and enthusiastic, decided to scrap all the proposals of the Committee for Native Education, and indeed, to scrap the Committee itself. In its place, he put down his own proposals and created a system which survives to the present day. This created a network of *zillah* schools, or district schools, and a central school and college in Madras (1841), which Elphinstone called the 'Madras University', but was later to be renamed the Presidency College of Madras. This remains modern Chennai's premier government college.

The Church Missionary Society was as active in the Arcot state (in modern Tamilnadu) as it was in Travancore. In the 1840s, the Secretary of the South Indian chapter of the Society was Rev. John Tucker, an Englishman from a rich and devout London family. His letters home struck a special chord in the heart of his eldest sister Sarah, a spinster who was confined to her room with a crippling disability, and lived only for her religious and missionary activities. Miss Sarah Tucker was determined to bring to the girls of south India the empowerment that comes with education and began to send large donations to the Society for this purpose. With this money, a small training school for primary teachers was founded at Kadachapuram, a small village near the undistinguished town of Sathankulam, in 1844. This school limped on with donations from Miss Tucker and her friends till 1857, when it was closed down following the sudden demise of its kindly patron. However, the benevolent friends of the late benevolent lady were determined that her dream should come true, and therefore, they raised enough funds to start the Normal School for Girls in Palayamkottai, a suburb of Tirunelveli, in 1858. This was converted into the Sarah Tucker College in 1895 and still runs by that name. And thus the name of the kind lady of a London suburb lives on in a town which she never saw in her life.

Other schools and colleges for women were to follow, including the Madras College for Women (1914), later renamed Queen Mary's College (1917). This was founded by Dorothy de la Hey, an Oxford graduate who had come to Madras to visit her brother, the Vice Principal of the 'Princes' College' in Madras. Her visit was indefinitely prolonged as she was persuaded to stay on to found the women's college and remain its Principal till her retirement 22 years later. It was, in fact, years before the College could be fully established. In the early days, there were no laboratories, and the girls had to travel to Presidency College to do their practical classes. But Miss de la Hey persevered, and today the institution she founded is autonomous, awarding its own degrees, including doctoral ones.

The Bombay Presidency was the third great swath of territory held by the East India Company, and in fact, it was in the west, at Surat in the modern Indian state of Gujarat, that the British had found their first foothold in India. However, the centre of British rule in the west soon moved to the new city of Bombay. This came up on an archipelago of seven little islands off the western coast of India, which had been seized by the Portuguese from the Sultan of Gujarat in 1534. The out-gunned Portuguese king later found it expedient to cede the islands to England, but as a face saver, he included it as part of the dowry of his daughter Catherine of Braganza, who was married to King Charles II of England in 1661. In 1686, the centre of the western operations of the East India Company was shifted from Surat to Bombay. By the middle of the eighteenth century the entire area around the islands was British territory.

Governor Gerald Aungier, who arrived in Bombay in 1669, and found it populated mostly by fisherfolk and subsistence farmers, determined to build it into the great *entrepôt* which it has subsequently become. It was he who proclaimed a policy of total religious toleration and offered various tax concessions to traders and skilled workmen. The result was that a motley collection of trading folk came flocking to the new city he was building. All these people were melded into a city by the British administration, creating India's most polyglot population and today's largest metropolis. Their burgeoning numbers could not fit into the few islands, and hence Governor William Hornby, who assumed office in 1782, took up a large scale scheme of land reclamation from the sea, which locally consisted mostly of shoals between the islands. This would eventually merge the islands into a single landmass, on which stands the mega-city of Bombay.

With such a patchwork of peoples, Bombay was, in many ways, a missionaries' delight. It was also a place where lonely European men could find solace by forming alliances with 'native' women who were not restricted by taboos. As early as 1718, the Rev. Richard Cobbe, chaplain to the East India Company's holdings in the Bombay island, decided to open a school for the education of the mixed-blood children of these alliances and give them 'a good Christian upbringing'. Cobbe's School, as it came to be known, was founded with one teacher and a dozen children. In fact, in 1815, under the influence of the Charter Act of 1813, the Europeans in Bombay founded the Bombay Educational Society and took over Cobbe's School. It was then taken over and expanded by the Christ Church at Byculla, and became the Christ Church School.

The real push for modern education in Bombay, however, came from a benevolent governor, Mountstuart Elphinstone (1779–1859). In 1819, after a long and honourable service with the East India Company, Elphinstone was appointed Lieutenant-Governor of Bombay, a post he held till 1827. In this position, Elphinstone proved to be one of the best rulers the entire colonial period ever threw up. Nevertheless, what Elphinstone is really remembered for is his great initiative to bring Western-style education to the 'natives'. Even before he became Governor, missionaries had founded the Education Society of Bombay in 1815, but it had been languishing for want of patronage. In 1820, with the weight of the new Governor's approval thrown behind it, the Society was able to attract rich donors like 'Nana' Jugonnath Shunkersheth, the Hindu jeweller, and Jumshethjee Jeejeebhoy and Framjee Cowasjee, the Parsee 'cotton kings'. A noted scholar named Mahomed Ibrahim Makhba also joined the Society with some of his associates. Elphinstone himself presided. The Society decide to open a subcommittee called the Native School and School Book Society, whose purpose was to raise funds from princes and rich merchants. With the Governor behind them, the Native School and School Book Society was permitted to start an English School at Bombay in 1824. It was only at the time of his departure from Bombay (1827), however, that Elphinstone's dream saw fruition. The notables of Bombay then got together to set up a college in his name, and thus was founded the Elphinstone College, in which was absorbed the English School. Classes at the college started in 1835. From 1856, the school and college became separate. A year later, when the authorities at Calcutta wanted to rename it as the Presidency College, the influential locals protested, for the memory of their benefactor was still strong with them. This is why Bombay has no Presidency College.

The missionaries, as in the other Presidencies, were emboldened by the changing climate in the Board of Directors to set up their own schools and colleges. The Scottish missionaries got in first, as usual. The Rev. John Wilson and his wife Margaret founded the Ambroli Church in Bombay (1831) near Girgaum, Chowpatty. Margaret Wilson had already founded a girls' school in 1829 for education in Marathi which later became the St. Columba's High School, one of modern Mumbai's prestigious schools for girls. In 1832, the couple founded the Ambroli English School, and a college wing was started in 1836, which was named the Free General Assembly's

Institution, in line with similar colleges in the other Presidencies. After several renamings, it came to be called the Wilson College, by which name it is known today. The college building, which is a major landmark in today's Mumbai, was designed in a florid mid-Victorian style by John Adams, a noted architect whose works dot the city today.

We have already met Fr. Henri Depelchin, the enterprising Jesuit priest who set up St. Xavier's College in Calcutta. In 1873, he was posted in Poona (Pune), when he got an urgent summons from the Apostolic Vicar of Bombay, Fr. Leo Meurin, who had established three small schools and a Catholic seminary in Bombay, but wanted to merge them into a college of repute. An able and distinguished Rector was needed for the new institution, and who would fit the bill better than Fr. Depelchin, with the laurels of the Calcutta college still fresh on his sacerdotal brow? And indeed, the now middle-aged priest rose to the occasion. Under his dynamic guidance, the amalgamated schools developed into the St. Xavier's College of Bombay—still one of the most iconic institutions in the city.

Seven years after the departure of Mountstuart Elphinstone, Bombay was again to get an experienced and dynamic Governor in Sir Robert Grant, son of the reformer Charles Grant. It was Grant, inspired by Bentinck's founding of a medical college at Calcutta, who suggested to the influential citizens of Bombay that their city should have the same (1835). But hardly had the permission to found a medical college come than Grant had a cerebral attack and died in 1838. There was an outburst of grief in Bombay, where the 'native' citizens mourned their late benefactor in a meeting at the town hall. Here a proposal was made by '*Nana*' Shunkersheth that Grant's dream project, the medical college, should indeed be founded and named after the late Governor. The proposal was carried with enthusiasm, and large donations were raised on the spot. It still took five years before the foundation stone of the Grant Medical College could be laid (1843), and classes began only in 1845.

Technical education came to Bombay much later than medical education. In the year 1887, when Queen Victoria celebrated her Golden Jubilee, i.e. fifty years on the throne, she had been officially Empress of India for ten years. The echoes of these celebrations have dissipated long ago, but they have left one tangible memory in Bombay. This is the Victoria Jubilee Technical Institute, or VJTI, an engineering college founded under the patronage of Lord Reay, the Governor of Bombay, and the enthusiastic support of another generation of prominent citizens like Dadabhai Naoroji, Pherozeshah Mehta, Dinshaw Wacha and the judges Mahadev Govind Ranade and Badruddin Tyabji. In 1998, it was ingeniously renamed the 'Veermata Jijabai Technical Institute', thereby erasing the colonial memory without changing the acronym.

Lord Reay's predecessor as Governor of Bombay was, rather unsurprisingly, another Scotsman, whose name was Sir James Fergusson. Under his patronage, a group of Marathi luminaries, founded the Deccan Education Society in 1884. This organisation, which still exists, founded more than a dozen colleges at Poona, Sangli and Bombay. Within a year of its inception, the Society founded the most famous of these, in Poona. It was named after its chief patron and most distinguished donor, the Governor himself, who gave 1,200 rupees to the cause, out of his personal savings.

The inauguration of the Fergusson College was carried out by William Wordsworth—not indeed the great poet himself, but his grandson, who happened to be the Principal of the Elphinstone College at that time. The College, which was founded in the same year as the Indian National Congress, grew apace with the political movement for India's freedom.

Women's education got off to a slow start in the Bombay Presidency. The Parsee community were perhaps the first to realise its importance, but given the prevailing prejudices of the time, their women were privately educated at home. These Parsee flowers were born to blush unseen, and we have little record of them, though surely they must have influenced their sons, some of whom went on to become great patriots and philanthropists. The missionaries did found quite a few schools for girls, as we have seen in the case of Mrs. Wilson. However, these were mostly attended by girls from Christian families and some from lower castes, and these subaltern women did not have much influence on the more powerful sections of society, which were dominated by upper caste Hindus and rich Muslims. For these communities, education of women was anathema and therefore rich Hindu and Muslim girls remained unlettered, essentially till the twentieth century.

However, the nineteenth century did see a great reforming couple, Jyotiba Govindrao Phule and his wife Savitribai, who tried to do for lower caste girls what Bethune and his associates had done for upper caste girls in Calcutta. In this context, we need mention only the girls' school for lower castes and 'untouchables' set up at Poona by the Phules in 1848, in the teeth of opposition by the influential Hindus of the city. By 1852, there were three such schools, with more than 250 girls enrolled. However, in 1858, in the wake of the disturbances of 1857, the tacit approval of the colonial masters was withdrawn, and all the Phule schools had to close down. In the decades following this setback, however, a few upper caste women became trailblazers for women's education in the Bombay Presidency. Chief among them were Sanskrit scholar and social reformer *Pandita* Ramabai Saraswati, and the brilliant Anandibai Joshi, who was the first Indian woman to acquire a medical degree, but alas! died at the tender age of 21. The clear statement of *Pandita* Ramabai aptly sums up the attitudes of the time: *"In ninety-nine cases out of a hundred the educated men of this country are opposed to female education and the proper position of women. If they observe the slightest fault, they magnify the grain of mustard-seed into a mountain, and try to ruin the character of a woman."* This sounds shockingly (but not surprisingly) relevant to our times as well.

It is possible, however, that some of the *Pandita*'s fiery words had reached the right ears, for when Justice Mahadev Govind Ranade and *Maharishi* Dhondo Keshav Karve started girls' schools in Poona in the 1880s, the opposition was more muted, and even a few girls from upper caste families began to trickle in. It was also Karve who later founded a university for women in Poona (1916), which was later shifted to Bombay (1920) and is now known as the *Shreemati* Nathibai Damodar Thackersey Women's University (SNDT Women's University for those who are short of breath).

The British rule in northern India was much more fragmented than in the other parts of the country, mainly because the north contained the seat of the waning Mughal power and its successor states. It was not till the Great Revolt of 1857 and its aftermath that the northern and most populous part of the subcontinent came under the complete sway of the foreign power. The political uncertainties of this period find their reflection in the growth of Western education.

Even today, despite its congestion and its squalor, its crumbling masonry and its ubiquitous zebu bulls, no one who has ever visited Benaras can fail to feel the irresistible charm of a city where three millennia coexist in perfect harmony. And so it was with British civil servant Jonathan Duncan—yet another Scotsman—who found himself posted as Governor of Benaras in 1788. Duncan immediately took steps to ban the doleful Hindu tradition of casting newborn female infants into the river Ganges. His efforts were met with protests and counter-protests. In dealing with these, he realised that some standardisation of Hindu traditions, including the preservation of Hindu scriptures and old manuscripts, was of primary importance. It was mainly with this end in view that he founded the Government Sanskrit College at Benaras in 1791. Thus was founded an institution which has remained the bulwark of Sanskritic education in India till today. The institution was renamed the Sampurnanand Sanskrit University in 1974.

Three hundred and thirty miles north-west of Benaras lies Agra, home to the most beautiful mausoleum in the world—the Taj Mahal. It is this glorious monument which makes the name of Agra known all over the globe and overshadows all the other achievements of this thriving city. Thus, few who are not from the city know that it houses the Agra College, one of the oldest colleges in India. For its foundation we must go back to *Pandit* Gangadhar Shastri, a renowned Sanskrit scholar and astrologer of the day, who was appointed to the court of the *Peshwa* Madhav Rao II, but remained stationed at Gwalior, where he served the Sindhia chieftain. *Pandit* Gangadhar lived a simple life and saved up his handsome salary as *raj-jyotisha* (court astrologer), which he bequeathed to found a college in his home town, which was Agra. These funds were used to start a Government College at Agra, better known as the Agra College, in 1823.

In the United Provinces of Agra and Oudh, which after Independence became the Uttar Pradesh, or northern state, the British had cantonments in every big city, but the biggest was in Cawnpore (Kanpur). It is still the biggest cantonment in modern India. With the soldiers came Army chaplains and then missionaries, Bible in hand, eager to win new territories for the Gospel of Christ. Chief among the latter were Anglican missionaries from the Society for the Propagation of the Gospel in Foreign Parts, or S.P.G. in short, which had been set up by a Royal Charter in 1701. As everywhere in the world, the cantonment in times of peace housed some hundreds of strong and energetic young men without very much to do except drink, play cards, and polish their accoutrements, and with money to spend in their pockets. Surrounding the encampment were villages where the people lived in ignorance and dire poverty. The natural result of all this was a growing number of children of mixed blood running in and around the cantonment, whose parentage was loosely acknowledged by their European fathers—at least until their transfer orders came. It was primarily to educate such children 'in the Christian way', that the S.P.G. priests at Cawnpore founded the Mission School in 1840, just as Cobbe had founded his school in Bombay more than a century ago. In the same year, services started at the Christ Church, a beautiful edifice which combines the Gothic and the Saracenic in a harmonious way. The Church was desecrated and damaged during the disturbances of 1857, and eventually handed over to the S.P.G. in 1862. They repaired and rebuilt the beautiful structure, and in 1866, celebrated its restoration by converting the Mission School into the Christ Church College. This still remains Kanpur's most iconic college.

The Rev. Midgley John Jennings was the priest who received Master Ram Chundra into the Anglican Church at Delhi. Jennings was Chaplain to the East India Company at Delhi, where, among other good works, he founded the S.P.G. High School in 1854. But he was not given long to nurture his creation. On May 11, 1857, the Mutiny began. With conspicuous courage, Jennings stayed back to comfort his flock, and thus his blood mingled with that of his parishioners when they were stoned to death on the streets of Delhi outside the Church of St. James. The church itself was gutted in the uprising. Four months later, the British soldiers retook Delhi and matched the rebels *sepoy*s savagery for savagery, washing out their own early humiliation in blood. However, by then the entire S.P.G. contingent of Delhi was extinct. In fact, the deep scars left by the Mutiny faded very slowly. It was only in 1877 that a new mission was founded at Cambridge University by the Rev. Edward Bickersteth of Pembroke College and set up in Delhi. These dedicated men soon came to be known as the Cambridge Brotherhood. It was this Brotherhood that, in 1881, re-established the school and college which the martyred Jennings had set up before his death. Not surprisingly, given the parallels between the fate of Jennings and that of the very first Christian martyr, they named it St. Stephen's College. Thus was founded one of modern India's best colleges. The aura of the ancient and the modern martyrs hovers over the institution, though few collegians today will remember the victims of 1857.

Unlike the other cities, the Catholic missionaries did not set up a college in Delhi during the early colonial period. It was only in 1919 that the Convent of Jesus and Mary, a girls school, was founded by the Religious of Jesus and Mary, a French order. However, the St. Stephen's College was followed in 1899 by the Hindu College of Delhi, founded by Indian nationalist Krishan Dassji Gurwale to provide '*non-elitist, non-sectarian and nationalist education to the youth*'. Like its counterparts elsewhere, the college was funded mostly by rich Indian philanthropists including trader *Rai Bahadur* Amba Prasad and the rich banker, *Rai Bahadur* Lala Sultan Singh. Then came the Ramjas College, founded in 1917. The founder, *Rai* Kedar Nath (1859–1942) was a rather remarkable person. Having served as judge under the British and saved some money, Kedar Nath resigned his position in 1911, and devoted the remaining thirty years of his life and all of his fortune to the cause of education. The Ramjas Foundation, named after his father *Lala* Ramjas Mal, was to fund several schools and the Ramjas College.

Almost a hundred miles to the north of Delhi lies the picturesque town of Roorkee, nestled at the foot of the mighty Himalayas. Like Cawnpore, this was a small village till the British established a cantonment there in 1953. Here, a unique educational institution was to come up. It was founded by noted civil servant Sir James Thomason, who had become Lieutenant Governor of the Northwest Province (1843). His crowning achievement was the foundation of India's first engineering college at Roorkee, which was intended, in the long run, to train Indians to undertake public works like laying roads and railways, and building bridges all over the subcontinent. Its immediate purpose, however, was to train Indian engineers to help build the Ganges canal, the pet project of Col. Proby Cautley to bring water to the parched land in the fertile *doab* between the Ganga and the Yamuna rivers. The canal was inaugurated in 1854, but by then, alas! the good-hearted Thomason was no more. To honour his memory, the opening of the Ganges canal was accompanied by the renaming of the engineering college as the Thomason College of Civil Engineering (1854). The college went from strength to strength, becoming the University of Roorkee in 1949 and the Indian Institute of Technology (IIT), Roorkee in 2001.

While the Ganges Canal and other irrigation works substantially helped the cause of Indian agriculture, during colonial times the crops and farming methods remained completely mediaeval. This did not escape the attention of one of British India's most dynamic—and most hated—Viceroys, namely George Nathaniel Curzon. It was 1905, and the bottom had just dropped out of the market for indigo dye. This bright blue natural colour was created in vast plantations in Bihar, which now became redundant because of the growing industrial synthesis of aniline—the active principle of indigo—from nitrobenzene in Germany, which made synthetic dye far cheaper. Facing ruin, the indigo planters were desperate to find other crops which could help them recover their losses, and found in the Viceroy a sympathetic ear. But Curzon's solution was characteristic of the man who made friends and enemies alike with every action—it was brilliant, it was forward-looking, and it pleased no one. He persuaded his rich American friend Henry Phipps Jr.—a partner of Andrew Carnegie in his steelmaking business—to donate the money to found an Agricultural Research Institute at Pusa, Bihar, close to the indigo-bearing areas. Though Curzon lured noted agricultural chemist John W. Leather to come out and guide this new Institute, the financial crises of the Indigo planters were much more immediate. One by one, the planters sold out and departed. The once-flourishing indigo plantations reverted to growing wheat and maize and subsistence farming for the hungry masses of Bihar. Only the Institute thrived.

In 1934, Bihar suffered a massive earthquake, and the Imperial Agricultural Institute at Pusa (for that is what it was called by then) was completely destroyed. By that time there was no real reason to continue the work in Pusa, which had sunk back into a rural area of no economic importance, and hence it was decided to shift the Institute to a suburb of New Delhi, which was renamed—no surprise—Pusa. The Institute has thrived there since then. It was here that Dr. M. S. Swaminathan (b. 1925), India's legendary agronomic scientist, worked on high-yielding hybrid varieties of wheat and rice, triggering the Green Revolution which made modern India self-sufficient in food despite her exploding population.

Four more colleges set up in the early days are the Khalsa College at Amritsar (1892), the St. Bede's College at Shimla (1904), the Robertson College at Jubbulpore (now Jabalpur) (1911) and the Hislop College at Nagpur (1846). With each of these colleges is associated a story, but those must be related elsewhere. For it is clear that we have come a long way from the pioneering efforts of some of the men and women mentioned in this section, and of many others who could not be mentioned in the interests of brevity. Today's India has nearly 40,000 colleges, which is still only

1 college for every 35,000 of her people. It is clear from this figure that universal education at the graduation level is still very much a pipe dream, so far as India is concerned. Nevertheless, it is undoubtedly a dream which needs to be dreamed.

3.8   Wood's Despatch and the Foundation of Universities

In the previous section we have encountered the names of various Universities. The foundation and growth of them will be discussed in this section. We may begin by noting that in the first half of the nineteenth century, though the English language and Western-style education had been introduced, the efforts were sporadic and much of it was connected with missionary activity, and some with counter-missionary efforts. Things were to change in 1854, at a time when Lord Dalhousie was governing the East India Company's domains. Two years earlier, the powerful position of President of the Board of Control—essentially the Cabinet Minister for India—went to a Whig liberal named Sir Charles Wood, later to be ennobled as Lord Halifax. Wood took up with dedication the task of determining the policies by which the Indian colonies would be governed, and in 1854, he sent to Dalhousie at Calcutta, a document which was of even greater impact that Macaulay's much-feted 'Minute on Education'. Entitled simply 'Despatch on Education', this document has come down to us as *Wood's Despatch*, a deceptively humble name which conceals the importance of what many historians have called the *Magna Carta* of western education in India.

What were the recommendations of this famed Despatch? It contained a whole blueprint for the setting up of a government-sponsored and government-regulated education system in India, which would impart to the 'natives' the best of western liberal education on the model of what was then current in England. In this sense, it systematised what Macaulay had wanted, and secularised what the missionaries were setting up. Wood's despatch clearly stated that it was the responsibility of the British government to provide modern, forward-looking education to the Indian masses. There were three stages of education envisaged in this visionary document. The first was *primary education*, to be imparted to children, essentially to teach them the three Rs, followed by *secondary education*, which was to impart high school level education as was given to boys and girls in the English 'grammar schools', and finally there was *undergraduate education*, which was to culminate in the creation of Indian graduates, who would be fit to man the vast Indian bureaucracy, the law courts and the engineering works across the subcontinent, as assistants to their British masters. This had, in fact, been Macaulay's dream. There was no need envisaged for *postgraduate education*, which, even in Britain, was limited only to those who could afford it.

The Despatch recommended the setting up of Departments of Public Instruction in five provinces, viz. Bengal, Madras, Bombay, the Punjab and the Northwest Frontier Province, each to be headed by a Director of Public Instruction (DPI). It would be the duty of these Departments to set up a wide network of educational institutions, i.e. schools, colleges and universities, and regulate the nature and standard of instruction to be imparted by them. At the highest level, three Universities were to be set up in the three Presidency towns, i.e. at Calcutta, Madras and Bombay, and were to be modelled on the University of London.[7] A network of degree colleges would then be set up, incorporating the existing ones, which would all be affiliated to these three Universities, which would then be the degree-granting institutions. Feeding these colleges would be a network of high schools, middle school and primary schools. It was recommended that professional and vocational education be given, and this led to the foundation of various 'polytechnics' to train professionals for technical work which did not require too much intellectual input. The education of women, just off to a shaky start in Victorian England, was strongly supported.

To man all these educational institutions, it was necessary to train teachers and keep them updated with modern developments, and hence Wood recommended the setting up of specific teacher's training institutes. It was also recommended that at the primary level, the education system could incorporate elements of traditional Indian teaching, including instruction in the vernacular, and the Government could help in printing textbooks and translating them into the vernacular. To support all this, Wood proposed the grant-in-aid system, whereby private institutions were encouraged to be set up with partial funding by the Government, and a system of scholarships for meritorious but needy students was added. Last, but not least, Sir Charles provided for a system of Government jobs for the educated Indian youth, which would require a major expansion of public works and Government initiative to ensure that the British youth seeking their fortune in India could continue to do so.

Wood's Despatch was a managerial *tour de force*, and one of the most influential documents ever drafted by a bureaucrat or a politician. It has been followed by several similar reports over the past century and a half, the latest being the National Education Policy proposed to and accepted by the Government of India as recently as 2020. Anyone who reads this latest report (or any of the previous ones) will be amused to see how little it differs in essentials from Wood's Despatch of 1854.

The Governor-General, Lord Dalhousie, to whom the Despatch was addressed, promptly wrote back to Wood, saying "*Your scheme is a very great one, and has been received with great applause in India.*" But Dalhousie was not able to implement much of it. The next Governor-General, Lord Canning, had hardly been in office a year when the Great Revolt of 1857 broke out. Nevertheless, Canning, a man with an amazingly cool head, proceeded to act on the provisions of Wood's Despatch while the Revolt was still raging. The three Universities envisaged in that document were all set up while fierce fighting was going on in northern and central India.

The University of Calcutta was founded by the passage of an Act in the Governor-General's Council on January 24, 1857. Justice Sir James William Colville, a noted scholar and the current President of the Asiatic Society was drafted as the first Vice Chancellor of the new University. The *Maharaja* of Darbhanga (in modern Bihar),

---

[7] The reason to choose the newly set-up University of London as a model rather than the famed Universities of Oxford and Cambridge (Wood himself read classics at Oxford) was because it had been set up in 1836 on purely secular lines, whereas the older institutions had grown out of mediaeval monasteries and still retained many mediaeval features, such as requiring the professors to be unmarried, a feature which lasted at Cambridge till 1860 and at Oxford till 1877.

Maheshwar Singh *Bahadur*, donated a plot of land opposite the College Square tank, and adjacent to the Hindu *Mahapathshala* to build the University. Even today the oldest University building, where the Vice Chancellor's office is located, is known as the Darbhanga Building. At the time of its foundation, all colleges in British India, from Lahore to Rangoon, and from Shimla to Ceylon, which were not in any of the other two Presidencies, were affiliated to the University of Calcutta. No University before or after has held sway over such a vast stretch of territory. Even today, there are about 160 colleges affiliated to this Unversity, in spite of a major decentralisation drive over the past two decades.

On July 18 of the same momentous year, the Council passed a similar Act, founding the University of Bombay. It started with just two faculties, one of Arts, which functioned from the Elphinstone College, and one of Medicine, which functioned from the Grant Medical College. The offices of the University were located at the Town Hall, now the iconic Asiatic Society building. We have already met with three of the prominent Indians who were named Fellows of the University. They were the tycoons Jagonnauth Shunkershet and Jamshedji Jeejeebhoy, and also Ramachandra Vithal '*Bhau Daji*' Lad, the last named being a medical doctor by profession, but also a polymath with formidable erudition on everything from *ayurveda* to numismatics. The other two were Bomanjee Hormarjee Wadia, another rich Parsee philanthropist, and Mahomed Yusoof Moorgay, the chief *qazi* of Bombay. The first Vice Chancellor was none other than the Rev. John Wilson, founder of the college that bears his name, but his formal appointment came only in 1868.

The University remained in its initial lodgings at the Town Hall, till the munificence of Sir Cowasjee Jehangir 'Readymoney', a rich banker, provided a lakh of rupees to build a proper home for the new institution. A plot of land was selected adjacent to the High Court of Bombay, backed on a road where once the ramparts of Bombay's Fort George ran. In 1868, the Governor and Chancellor of the University, Sir William Fitz-Gerald, laid the foundation stone of the new campus, which was then built to a magnificent design by famed Victorian architect Sir Gilbert Scott. A clock tower of matching opulence was subsequently built with a donation from 'cotton king', Sir Premchand Roychand. It still carries the name of his blind mother, Rajabai.

The last of the three was the University of Madras, founded by another Act passed on September 5, 1857. Initially, it operated out of the Presidency College, with Sir Christopher Rawlinson, the then Chief Judge of the Madras High Court, functioning as Vice Chancellor. Rawlinson retired, however, in 1859, and was replaced by Sir Walter Elliot, a civil servant who was also an archaeologist, linguist, naturalist and numismatist. It was under him that the University began to take shape. The gorgeous Senate House of Madras University was built between 1874 and 79 by Robert Chisholm, a pioneer of the Indo-Saracenic style. Today the University is spread over six different campuses and has 87 academic departments, as well as 121 affiliated colleges and 53 research institutes.

The Aligarh Muslim University was founded in 1875. The story of its origins are related in the next section. In 1882 came the Punjab University and in 1997 came the Allahabad University. The University of the Punjab was founded on the initiative of a remarkable personality, who is all but forgotten today. He was Gottlieb Wilhelm Leitner (1840–1912), born into a Hungarian Jewish family, who was a linguist *par excellence*, able to speak some fifty different languages. He became a Professor of Turkish and Arabian at King's College, London at the age of 23. Using his ability to speak these languages, and adopting the Muslim name of Abdur Rashid Sayyah, Leitner explored the then-inaccessible regions of Dardistan, mostly now in northern Pakistan, and wrote a magisterial *History of Islam*. A formidable scholar, Leitner was appointed Principal of the Government College at Lahore in 1864. The University of the Punjab came into being one year after his departure from India, but bore the stamp of his plans and ideas. After the partition of India in 1947, the University split into two. The original University remained at Lahore, in the new state of Pakistan, while the professors who moved to India created the Panjab University, with a slightly different spelling to indicate the distinction. This moved to its new campus at Chandigarh in 1956.

The city of Allahabad, now renamed Prayagraj, lies at the confluence of the sacred river Ganges with its biggest tributary, the Yamuna. It has been revered as a spot of exceptional holiness by Hindus down the centuries. The British built a huge cantonment there, and set up a new city along colonial lines. It was here that India's fifth University (sixth, if we count the Danish one at Serampore) came up in 1887. However, the story goes back to 1873, when an initiative taken by Sir William Muir, the Lt. Governor of the United Provinces, lead to the foundation of the Muir Central College, as it was called. It was funded by generous contributions from several quarters, mostly rich landowners, which included the hereditary *Maharaja* of Kashi. The foundation stone of the new college was laid by the reforming Viceroy, Lord Northbrook. The opulent University buildings were designed in Indo-Saracenic style by William Emerson, who is more famous for having designed the iconic Victoria Memorial in Calcutta. The University grew to be a centre of excellence and was, for a time, known as the 'Oxford of the East'.

In the early years of the twentieth century, more Universities sprung up at Benaras, Mysore, Patna, Hyderabad, Lucknow, Santiniketan and Delhi. The Benaras Hindu University, or BHU, as it is generally known, was set up by the personal efforts of Hindu reformer *Pandit* Madan Mohan Malaviya. The Visva-Bharati, a unique institution founded on the traditions of the ancient Indian *gurukul*, was founded by Nobel Prize-winning poet Rabindranath Tagore. The University of Delhi was set up with Dr. Hari Singh Gour, a noted jurist, humanist and poet, as its first Vice Chancellor. Like the other Universities, it was initially just a degree-granting body. In 1937, however, it received a dynamic Vice Chancellor in Sir Maurice Gwyer, who actively set up postgraduate Departments in Arts, Sciences and Law. Gwyer, who continued to serve as Vice Chancellor till his retirement in 1950 (aged 72), is credited with being the 'maker of Delhi University'.

With the proliferation of Universities in India, science education, and the beginnings of scientific research spread across the country. Today, modern India has more than 800 universities, but that is still around 50 colleges to every University, and one University for around 1.75 *million* people. Clearly, the work of the early educationists in India is very far from complete.

3.9   Muslims Join in

Though the adoption of Western scientific thinking was initially limited to the upper-caste Hindu *bhadralok*, it was not very long before enlightened Muslims began to appreciate the necessity for imbibing some of these new-fangled ideas. Prominent among them was Sir Syed Ahmed Khan. Like Ram Mohun Roy, he was a servant of the Company *bahadur* and even rose to become a minor judge. It is related that in 1855 this versatile scion of a long line of Mughal *umrah* created a magnificent scholarly Urdu[8] edition of the *Ain-i-Akbari* of Abu'l Fazl, the Mughal Emperor Akbar's erudite minister. Sir Syed wanted the famous poet *Mirza* Ghalib to write a *taqreez*, i.e. a short laudatory poem introducing the work. The poet complied, but instead of the customary praise, his lines were heavily critical of Sir Syed's looking backwards to the days of Akbar and strongly advocated that he stop wasting his time on the dead past and concentrate on the new *ain*[9] of the *sahib*s. It was, apparently, this rebuke which galvanised Sir Syed into action to spread Western education among his co-religionists. It makes a pretty story that a consciously world-weary inheritor of an ancient tradition of poetry revelling in wine, women and double entendres should issue such a forthright reproof to a savant who had been writing on everything from Newtonian mechanics to English law for some thirty years. But of its effectiveness there can be no doubt.

In the wake of Ghalib's rebuke, Sir Syed Ahmed Khan established the Gulshan School at Moradabad (1859), the Victoria School at Ghazipur (1863) and eventually the Muhammedan Anglo-Oriental College (1875), which became the Aligarh Muslim University in 1920. A British loyalist like Ram Mohun Roy and Ram Chundra Lall, he was posted in Ghazipur when he founded the Scientific Society of Ghazipur, with the help of one *Raja* Jai Kishen Das Chaube (*sic*), who remained his close associate for life. When Sir Syed was transferred to Aligarh in 1864, the Society moved with him and became the Scientific Society of Aligarh. The Society then created the Aligarh Institute and a building was constructed with a Reading Room, Library, Laboratory, Museum and Lecture Hall. In 1867, Sir Syed was transferred to Benaras, but the Society remained in Aligarh, in the safe hands of *Raja* Jai Kishen. By 1867, however, the entire effort of Sir Syed and his friends was taken up by the Angle-Oriental College, and the now-defunct Society was given a quiet burial. During its decade or so of life, the Society brought out a paper called the *Aligarh Institute Gazette*, which discussed multifarious topics including scientific ones like the solar system, plant and animal life, human evolution, etc. This legacy of science was continued by the Anglo-Oriental College and is being continued by the Aligarh Muslim University.

A similar organisation grew up at Muzaffarpur in Bihar, but with somewhat different origins. Bihar had borne the brunt of the oppression of European indigo and opium planters, with the full connivance of the British courts of law. It was a Subordinate Judge at the *Sadar Amin* court at Muzaffarpur, Syed Imdad Ali Shah, who was so distressed by these goings-on that he founded, in 1868, a 'British Indian

---

[8] The original was in Persian, the official language of the Mughal courts.

[9] Meaning 'ways'.

Association', with the intention of *'criticizing the proceedings of the government, and defending the people from oppression by conveying their true complaints to the government'*. Conditions, however, had improved somewhat in the wake of the Indigo Commission of 1860, and soon the efforts of Syed Imdad were channelled in a different direction. The 'British Indian Association' metamorphosed into the 'Bihar Scientific Society' in 1872. The mandate of the Society was to bring European ideas in the intellectual, social and moral spheres to the 'natives' of Bihar, especially the Muslims. Like the Delhi and Aligarh groups, they brought out a periodical called *Akhbar-ul-Akhyar* and organised a series of meetings called *Anjuman-e-Tehzeeb*. We learn from Garcin de Tassy, a French orientalist who was an enthusiastic member of the Society, that they translated books on Trigonometry, Optics, Physiology, Algebra, Mechanics, Philosophy, History, Agricultural Sciences and Masonry, as well as the *Materia Medica*, from English into Urdu and Hindi.

But the energetic Syed went further. He persuaded the *Bhumihar-Brahman Sabha*, an association of rich landowners founded by the *Raja* of Benaras (1889) to fund the setting up of a college at Muzaffarpur. This 'Bhumihar-Brahmin College', as it was originally known, was set up in 1899, and after various renamings, survives today as the Langat Singh College—often shortened to L.S. College. In the first meeting of the *Bhumihar-Brahmin Sabha*, where initially only 5,000 rupees were promised for the College, *Babu* Langat Singh had startled his cautious fellow-members by promising to contribute 125,000 rupees, thus shaming them into donating 50,000 rupees on the spot. It is fitting, therefore, that the College should be named after him today.

As for Syed Imdad Ali, he was transferred to Gaya, where he founded a branch of the Muzaffarpur Society and continued to spread the light. After retiring from the Judicial Service, he settled down in Bhagalpur, where he died in 1886, in his mid-seventies. Today one searches the L.S. College website in vain for his name.

3.10 Cultivation of the Sciences

We now come back to Calcutta, and relate the story of the foundation of one of the most influential of the science societies set up in colonial India. This is intimately connected with the story of one of the most paradoxical characters of the time. Mahendra Lal Sarkar (1833–1904) was a poor orphan brought up by his uncles in Calcutta. Educated as a free student at Hare School, he joined the Hindoo College at the young age of 16 on a scholarship and imbibed all the Science that he could during the five years he spent there. However, influenced perhaps by the early death of his parents, he was determined to study medicine, and hence, in 1854, he got admitted again to the Calcutta Medical College. Here he was very well thought of by his teachers, and graduated with flying colours in medicine, surgery and midwifery (1860). Three years later he was awarded an M.D. (one of the first three M.D.s from Calcutta) and set up in practice as a fashionable young doctor. Soon he acquired a great reputation—so much so that other doctors in practice would send him their difficult cases. In 1863, he was elected Secretary of the Bengal branch of the British Medical Association.

Till this point, there was nothing in the career of the young Mahendra Lal Sarkar to distinguish him from other eminent Bengali doctors who came later like Nil Ratan Sircar, Radha Gobinda Kar or Bidhan Chandra Roy. However, at this stage of his career, Mahendra Lal became disillusioned with the medical practices of his time. Instead, he became interested in homoeopathy. Those who think of homoeopathy as a pseudo-science and even a system of quackery may stand aghast at this change in the young physician, but it must be remembered that in the mid-nineteenth century, the standard allopathic practices themselves included a good deal of quackery, including bloodletting,[10] mesmerism and electric shock therapy. Surgery in those days was only just a little better than butchery. It was done without anaesthetics, antiseptics or antibiotics—the three essentials of modern surgical practice. As a result those patients who survived the pain and trauma often died of post-operative infection. Blood transfusions were generally deadly, since blood groups had not been discovered, and most often the patient died of haemolytic shock. Child-bearing carried such a risk of death that the expecting mother was given a *saadh*, i.e. a meal of her favourite dishes—in the same spirit as a last meal is given to a convict before execution. It is not surprising, therefore, that the sensitive mind of Mahendra Lal Sircar was impressed by the simple and non-invasive methods of homoeopathy, and its declared aim of stimulating the body to cure itself.

Incidentally, during his allopath days, Mahendra Lal, like his fellow allopaths, was prone to dismiss homoeopathy as the practice of charlatans and quacks. In 1864, he was handed a copy of a new book called *The Philosophy of Homoeopathy* by William Morgan, and asked to review it for a journal called *Indian Field*. Determined to turn the full force of his invective on this upstart philosopher, Mahenda Lal Sarkar started reading the book with grim determination. Soon, however, it seemed to him that the arguments made in the book were rather compelling, and should not be dismissed without scientific tests. This drove him to seek clarifications from *Babu* Rajendra Lal Dutta (1818–1889), Calcutta's first Indian practitioner of homoeopathy. This merchant prince, who was also a medical man and a staunch Derozian, had started practising homoeopathy and was credited with curing *Pandit* Ishwar Chandra Vidyasagar of migraine, and *Raja* Radhakanta Deb of a gangrenous foot. In both cases, it is related that homoeopathy had worked when allopathic treatment had failed. One does not know what transpired in the discussions between the two savants, but Mahendra Lal emerged from it a clear convert to the philosophy of homoeopathy. It is reported that the good doctor had prepared homoeopathic medicines himself, tried them on patients, and found them to be efficacious. Without knowing more details, one cannot comment further on Mahendra Lal's tests, but what we do know for certain is that *he* believed that homoeopathy works.

---

[10] Bloodletting was once believed to be a cure for almost all diseases, including (according to an eighteenth century text-book) acne, asthma, cancer, cholera, coma, convulsions, diabetes, epilepsy, gangrene, gout, herpes, indigestion, insanity, jaundice, leprosy, ophthalmia, plague, pneumonia, scurvy, smallpox, stroke, tetanus, tuberculosis, and many others.

The new acolyte was now ready to spread the word. On February 16, 1867, at a meeting of the Medical Association—of which he had by now become the Vice-President—Mahendra Lal Sircar delivered a stirring address with the innocuous-sounding title *"The uncertainties in medical sciences and the relationship between diseases and their remedial agents"*. The subject of the lecture, however, was the inadequacy of allopathic medicine and the superior efficacy of homoeopathy. As he spoke, his audience of British and Indian doctors erupted with indignation and the lecture ended in an uproar. The apostate was forced to resign his Vice-Presidency and soon, to quit the BMA altogether. He also found his extensive practice shrink to a few loyal friends and patients.

Nothing daunted, Mahendra Lal founded a new practice, this time as a homoeo-pathic doctor. In the beginning, he had few patients, and had all the time in the world to start a new journal (1868)—the *Calcutta Journal of Medicine*—in which he tried to explain his reasons for promoting homoeopathy. *Babu* Rajendra Lal had to help his new protégé by redirecting some of his own patients to the young practitioner. Soon, however, his clinical instincts appear to have proved as powerful in this new method as in the old one, and Mahendra Lal began to acquire an increasing reputation as a worker of wondrous cures. Eventually he became a household name in Calcutta.

However, in the midst of all his successes, Mahendra Lal never forgot the brick-bats he had faced from the medical establishment in 1867. His magnanimous mind, however, felt no bitterness against his former colleagues, but instead descried an acute need for a truly scientific way of thinking among his fellow men. The question which puzzled him most was—why would not these otherwise reasonable doctors listen to his arguments and look at his results before castigating him as a charlatan? And the answer that came to his mind was that this was because their minds were closed i.e. they were *practitioners* rather than originators of science. It was, thus, to counter what he felt was professional narrow-mindedness that in 1876, Mahendra Lal founded the *Indian Association for the Cultivation of Sciences* (IACS), an organ-isation which still stands forth as one of modern India's leading research institutes. Writing in his Calcutta Journal, he explained that his new institution would function on the model of the Royal Society and the British Association for the Advancement of Science, but *"We want a different institution altogether. We want an institution which shall be for the instruction of the masses, where lectures on scientific subjects will be systematically delivered and not only illustrative experiments performed by the lecturer, but the audience should be invited and taught to perform themselves. And we wish that this institution be entirely under native management and control."*

The first donation of 1,000 rupees (about Rs. 150,000 in today's denomina-tion) came from *Raja* Joy Kissen Mukherjee of Uttarpara in 1870 and was followed by a flood of contributions from, among others, the *Maharaja* of Patiala, *Maha-rani* Swarnamoyee of Cossimbazar, the *Maharaja* of Cooch Behar, *Raja* Kamal Krishna Deb of Sovabazar, *Babu* Kali Krishna Tagore of Posta and Justice Sir Romesh Chandra Mitter of the Calcutta High Court. Mahendra Lal himself contributed 1,000 rupees to his fledgling institution. A house was rented at 210 Bowbazar Street, and here, on July 29, 1876, the Association was formally inaugurated. In the begin-ning the Association was more-or-less what its name suggests—a debating society, where cultured men of leisure would gather for an evening of lectures on the new sciences, and a sumptuous dinner at the end of it. Mahendra Lal was the most prolific lecturer. In the five years from 1878 to 1883, he delivered no less than 154 lectures on subjects as diverse as electricity, magnetism, heat, light and sound, to say nothing of medical topics. Other prominent lecturers included *Acharya* Sir Jagadis Chunder Bose (inventor of wireless telegraphy and pioneering biophysicist), *Acharya* Prafulla Chandra Ray (pioneering chemist), *Rai Bahadur* Chunilal Bose (toxicologist) and Sir Asutosh Mukherjee (mathematician and jurist). It was, however, Bose who insisted on holding practical classes as well as demonstrations in the lectures, to give the audience the true feel of empirical Science.

In 1891, Mahendra Lal was able to persuade the philanthropist Ananda Gajapati Raju, *Maharaja* of Vizianagaram, to donate 40,000 rupees for the building of a new laboratory for the Association. The Vizianagaram Laboratory, as it was called, came up in the premises of the Association at 210 Bowbazar Street. However, it was used more for demonstrations than for actual research. It was only three years after Mahendra Lal's death that a young Indian accountant, working for the British Government, applied to the new Secretary of the Association for permission to use the Vizianagaram Laboratory for some scientific researches he had in mind. This Secretary[11] was only too pleased to permit this, since the expensive equipment in the laboratory was lying largely unused. The youthful accountant was C.V. Raman, and it was at 210 Bowbazar Street that Raman and his student K.S. Krishnan discovered what is known as Raman Scattering. The Nobel Prize followed. Eventually the Indian Association moved to new premises of its own in Jadavpur, then a suburb of the city, and it continues to stand there as India's oldest research institute.

---

[11] It was Amrita Lal Sircar, son of Mahendra Lal Sircar, who succeeded his father as IACS Secretary and carried on his mission with great enthusiasm.

Chapter 4
Indian Science Comes into its Own

Towards the close of the nineteenth century, the diligent efforts of educators in India began to yield their first fruits, producing a small number of native-born scientists who could equal the best in the world. They were not isolated geniuses like Radhanath Sickdher or Master Ram Chundra, but grew out of the intellectual milieu created by the proliferation of Western education across the country. To do justice to these pioneers would require a full volume—in fact, several volumes—but the story of particle physics cannot commence without a brief introduction to these savants and their contributions. This chapter is, thus, an attempt to condense their achievements into a few short pages.

Medicine was, perhaps, the first area in which the new learning made its most visible impact. Though it is not central to the story, the tale of Mahendra Lal Sircar illustrates the influence of medicine on other branches of science. In fact, despite the survival of some mediaeval practices such as described in the previous chapter, the nineteenth century saw rapid advances which took Western medical science far ahead of what had been achieved in the previous few thousand years. In particular, Western medical practitioners broke away from the old pattern of thought where disease was attributed to imbalances in the four 'humours', and increasingly embraced the 'germ theory' propounded by Louis Pasteur and his followers. Antiseptics saved thousands of lives, and anaesthetics brought immense relief from suffering. One by one, the culprits for the great scourges of humankind—plague, cholera, typhoid, diphtheria, typhus, malaria, syphilis and tuberculosis—were identified. A whole new pharmacopoeia was created as physicians sought 'magic bullets' to cure these specific diseases, and culminated with the discovery of antibiotics in the next century. Instruments like the thermometer, the stethoscope and the microscope became invaluable aids for diagnosis. More than ever, it became apparent that the 'art of healing' was really the *science* of healing, and that it had intimate links with the other sciences. The final nail in the coffin of the old school of thought was the theory of evolution, which, from the medical point of view, meant that tremendous insights could be

S. Raychaudhuri, *The Roots and Development of Particle Physics in India*,
SpringerBriefs in History of Science and Technology,
https://doi.org/10.1007/978-3-030-80306-3_4

gained into the working of the human body by studying animals. Medical science was truly on a roll.

We have already noted that Western medicine was able to effect cures where the traditional Indian methods failed and this was perhaps the single greatest factor in reconciling the mass of Indians to the fact that Western science represented progress. Almost an equally important role was played by the introduction of railways into India. Initially reluctant to travel by train for fear of losing caste, Indians were soon to throng these iron carriages, and still do so in enormous numbers.[1] The general Indian reaction to railways was first one of hostility, then of scepticism, then of awe mixed with envy, then of acceptance, followed by enthusiasm, and finally, of taking the railways for granted. And this pattern was repeated for much of Western science, among other things foreign.

4.1   The 'Master of Nitrites'

Our story now turns to a man who lived 82 years, but whose influence lives on a century and a half after his birth. This was Prafulla Chandra Ray, the chemist, who was honoured by the British with a knighthood of the smallest denomination, and by his own countrymen with the supremely respectful title of *Acharya*. However, Ray was not just a chemist, but also a scholar and an entrepreneur. He has justly been called the 'Father of Modern Indian Chemistry'.

Ray was born in 1861, to a wealthy *zamindar* family in the village of Raruli, now in Bangladesh. As a child, Prafulla Chandra was educated at the village school founded by his father *Babu* Harish Chandra Ray, a liberal-minded philanthropist. From his youth, Prafulla Chandra was a voracious reader and by the time he was 12 years old, he had devoured both English and Bengali classics and historical works, in addition to learning Latin and Greek. He passed the Entrance Examination in 1879 and was admitted to the Metropolitan Institution,[2] founded by another Bengali stalwart, *Pandit* Ishwar Chandra Vidyasagar. From here, he graduated in 1881, with a First Arts Diploma from the University of Calcutta. Among the compulsory subjects in the First Arts class were Physics, Chemistry, and Mathematics, for which Prafulla Chandra attended classes at the Presidency College as an external student. Chemistry, in particular, was taught very ably at the Presidency College by Professor (later Sir) Alexander Pedler. This inspiring teacher was a first class empiricist and his classroom demonstrations of chemistry experiments had a profound influence on the young Prafulla Chandra.

After graduating from the Metropolitan Institution in 1881, Prafulla Chandra was determined to fulfil his father's dream of sending his brilliant son to Cambridge University to study. Unfortunately, *Babu* Harish Chandra Ray had squandered his

---

[1] In 2019, the Indian Railways transported more than 23 million passengers *per day*, roughly three times the entire population of Switzerland.

[2] Renamed the Vidyasagar College in 1917, after the passing of its great founder.

fortune and could not afford it any more. However, there did exist a window of opportunity for meritorious students. The Gilchrist Educational Trust, in 1881, was offering two scholarships for Indians to go to the University of Edinburgh for higher studies. Prafulla Chandra resolved to appear for the selection examination, and won one of them. And so, with stars in his eyes, the young man from Raruli village sailed for Britain on the S.S. *California* in 1882.

At Edinburgh, Prafulla Chandra had the good fortune to be taught by Alexander Crum Brown, the noted organic chemist whose method of drawing diagrams to represent molecules (circles for atoms and lines for valence bonds) is universally used today. In 1885, he obtained the B.Sc. degree (then equivalent to a Masters) and spent the next two years in research, as the terms of the scholarship allowed him to do. He worked nominally under Crum Brown but did not feel inclined to work in organic chemistry, which was the learned Professor's interest. Instead, he chose to work more-or-less independently in the area of inorganic chemistry. Within two years, he had completed his thesis, where he established that there appear to be a class of compounds which are 'double-double sulphates'. The correctness of his empirical discoveries has never been disputed, though the modern interpretation has changed from interpreting these as compounds to mixed crystals of two kinds of double sulphates. For this work, in 1887, Prafulla Chandra Ray was awarded the D.Sc. degree of the University of Edinburgh—the first Indian to be awarded a doctorate in Chemistry—as well as a Hope Prize Scholarship which enabled him to spend another year in Britain.

Returning to India in 1888, with a great reputation, and a letter of recommendation from Crum Brown in his portfolio, the young scientist still found himself unable to get a job. The only decent salaries were to be had in the Indian Educational Service (IES), run by the British Government, but there was a definite glass ceiling, for not only were Indian lecturers few and far apart, but they were also put in a special 'Provincial' category, where they had to accept two-thirds of the salary given to their white-skinned peers. There was only one college in the whole of India where a proper chemistry course was taught, and that was in Presidency College, where the indefatigable Alexander Pedler was still teaching with all his old verve and vim. It took, however, fully a year of lobbying before Prafulla Chandra could get a *temporary* position as *Assistant* Professor in Presidency College. After joining Presidency College, though, Prafulla Chandra made a name for himself as a teacher— not a stately and awe-inspiring figure like Professor Pedler—but a more down-to-earth person who would crack jokes in class and quote at random from the poems of Rabindranath Tagore and the *shloka*s of the ancient Indian chemist Nagarjuna.

Now settled in a job, Prafulla Chandra wanted to set up his own laboratory, but it was not till 1894 that a proper research laboratory in Chemistry was inaugurated at the Presidency College. While the new laboratory was being set up, Prafulla Chandra carried out researches in the teaching laboratory after classes were over. The topic he chose was vastly different from what he had done in Edinburgh and much more in consonance with his avowedly Indian roots. It was a study of adulteration and setting up of standards for mustard oil and *ghee* (clarified butter), the major mediums used in Indian cooking—at least in the northern half of the subcontinent. One shudders to think what a similar investigation might show up in the more unscrupulous world of today.

Once the new laboratory was set up, however, Prafulla Chandra went right back to his forte, which was as an inorganic chemist. One of Prafulla Chandra's experiments involved converting elemental mercury to mercurous chloride (Hg_2Cl_2) or calomel—commonly used as a reagent as well as a medicine in those days. An intermediate step in this was to convert elemental mercury (Hg) to unstable mercurous nitrite—$Hg_2(NO_2)_2$—which could be chlorinated to yield calomel. As mercury in its mercurous or mercury (I) state is rather unstable, tending to decompose into the mercuric or mercury (II) state and elemental mercury, and the nitrite radical $(NO_2)_2$ is also notoriously unstable, tending to split into a pair of more common nitrate (NO_2) radicals, mercurous nitrite was thought to be doubly unstable, and hence an appropriate intermediate state in the synthesis of stable compounds like calomel. In a delicate experiment, requiring more than usual dexterity, Prafulla Chandra washed metallic mercury with a dilute nitric acid solution, and to his surprise, found that he was able to precipitate stable yellow crystals of a hitherto unknown compound. Though Prafulla Chandra's initial guess was that this was some basic salt of mercury, chemical tests soon established that these crystals were those of mercurous nitrite—the impossible compound, where both radicals were unstable, but seemed to come together in a stable form. This startling discovery was presented to the Asiatic Society in a lecture by Prafulla Chandra in December 1895, and published in their Journal in 1896. The prestigious *Nature* magazine reported on this work later in the same year, and Prafulla Chandra's international reputation was made.

The discovery of mercurous nitrite spurred Prafulla Chandra and his students on to the study of a whole family of nitrates, nitrites and hyponitrites—a programme which kept them busy for a couple of decades. But it was nitrites which were the centre of this research, from nitrites of weakly reacting elements like gold, silver and platinum, to the commercially-useful ammonium nitrite, used to make explosives and fertilisers. At the Chemical Society of London, Prafulla Chandra was hailed by Sir William Ramsey (discoverer of the noble gases) as the '*master of nitrites*'.

If we look at the 158 publications of Prafulla Chandra Ray's long career, we will find that about 90 of them deal with nitrogen compounds of some kind—mostly nitrites—and of the remaining 68, around 40 of them deal with sulphur compounds—sulphides, sulphates and sulphites—showing that at a later stage, having conquered the world of nitrogen, Prafulla Chandra returned to his first love, i.e. sulphur compounds. Dispersed among these we find some general articles on topics like valence and one solitary review article on ancient Hindu chemistry. However this last was a major passion with the man soon to be acclaimed as *Acharya*, and it is to this side of his multi-faceted personality that we must now turn. For it is at this point that he began to write his monumental *History of Hindu Chemistry*, of which Volume I appeared in 1903 and Volume II in 1909. It is amazing that this fantastic work of scholarship, which brought forth much lost knowledge from long-forgotten ancient manuscripts, was carried out *in parallel* with Prafulla Chandra Ray's path-breaking work in nitrite research. Such an enormous burden of work would have affected the quality of the product in any normal individual, but it was not so with this man. If he did something, he would do it one hundred percent. And so, he set out, in his dogged fashion, to procure as many manuscripts and read up as much as he could on the subject of Indian chemistry in the ancient and mediaeval periods. The result was the two tomes of the *History*. He writes with pride that this lead to the great Berthelot hailing him as a *savant*—a description which was more appropriate than the French chemist might have thought. Incidentally, while extolling the advancements in ancient Indian chemistry, Prafulla Chandra did not hesitate to criticise the ancients on several counts, including the excessive and barren use of gold and mercury, and the fact that they were never able to demarcate the boundaries of chemistry, medicine and magic.[3]

Prafulla Chandra's interest in ancient Hindu chemistry may have originated in his Sanskrit studies and may also have been influenced by his nationalistic urges. It should be noted that in this context 'Hindu chemistry' means simply 'Indian chemistry', the very word Hindu being unknown in the subcontinent before the advent of the Arabs, as described in the Prologue. Let it be said at this point that Prafulla Chandra's historical vision was severely objective. In his 1886 *Essay on India*, he had written "*We find there is a tendency among a certain class of writers to single out some of the worst types of Mahomedan* (sic) *despots and bigots, and institute a comparison between the India under them and the India of to-day... It is forgotten that at the time when a Queen of England was flinging into flames and hurling into dungeons those of her own subjects who had the misfortune to differ from her on dogmatic niceties, the great Mogul* (sic) *Akbar had proclaimed the principles of universal toleration... Religious toleration, backed by a policy dictated no less by generosity than by prudence, was the rule and not the exception with the Mogul emperors.*" These wise words and clear vision have not lost their relevance even today.

Remarkable as they may seem, if Prafulla Chandra's achievements were confined to those related above, he would have appeared only as a footnote in histories of chemistry, or his *History* might be quoted once in a while as a reference work. The real reasons for remembering him must now be told.

The first of these reasons is the fact that Prafulla Chandra created—almost single-handedly—the first science-based industry in India. During his enforced year of idleness after returning from Britain, he had acquired some knowledge of botany and zoology, and had been impressed by the '*thousand and one raw products which Nature in her bounty has scattered broadcast in Bengal.*' Ever after, Prafulla Chandra would pursue the cheap manufacture of chemicals using local resources. He turned his hand to phosphates of soda, which could be cheaply manufactured by calcining bones. Using his rented house at 91 Upper Circular Road, in Calcutta as a manufactory, Prafulla Chandra began to purchase huge quantities of bones from butchers and leave them on his terrace to dry out. Not surprisingly, the smell of putrefaction and the flocks of crows this attracted got the young scientist into trouble with his

---

[3] A criticism which may equally be levelled at some of the great mediaeval European alchemists like Albertus Magnus, Paracelsus, Roger Bacon, Ortolanus, Cornelius Agrippa and John Dee, to mention just a few.

genteel neighbours, and soon he had to move his works to a less upmarket suburb near Maniktola (now in the heart of the city). Here, however, the smell of bones being burnt into charcoal attracted the unwelcome attention of the police, who thought it was a human body being secretly cremated and sternly accused the young professor of murder! It was among such trials and tribulations that the Bengal Chemical and Pharmaceutical Works was born.[4] The story of the nascent works and its success in later years is too long to tell here, but it may suffice to say that it still stands and some of its products like *Aqua Ptychotis*, *Pheneol* and *Cantharidine Hair Oil* have become household words, especially among Bengalis.

The final—and perhaps the most important—part of Prafulla Chandra's rich legacy is the fact that this amazing man not only studied and carried out research in chemistry and the history of chemistry as well as founding an industry, but that he built a school of first class chemists, who went on to create much of modern Indian chemistry as it stands today. A pleasing anecdote is told of some *memsahib* who asked Prafulla Chandra if he had any children, to which the life-long bachelor replied that he had seventy-three, shocking the uptight lady, but meaning, of course, the students who would proliferate his intellectual legacy rather than his blood. Indeed, the list of Ray's students and collaborators is very long, but it should suffice to mention just a few of these giants: Nil Ratan Dhar (1892–1986), founder of physico-chemical research in India, Sir Jnan Chandra Ghosh (1894–1959), whose pioneering work on electrolytes was instrumental in the development of the Debye-Hückel theory, and who later brought colloid chemistry to India, Jnanendra Nath Mukherjee (1893–1983), another pioneer in colloid chemistry and soil chemistry, and Bires Chandra Guha (1904–62), known as the 'Father of Biochemistry in India'. Others, who did not achieve so high a level of fame, were excellent scientists nevertheless. Between them, this galaxy of Prafulla Chandra's bright students carried on the tradition set by the Master and erected the proud edifice of Indian chemistry, pharmaceuticals and science-based entrepreneurship, which is one of the engines on which today's upwardly-ambitious nation hopes to propel itself to superpower status.

In 1912, the reforming Vice-Chancellor of Calcutta University—the redoubtable Sir Asutosh Mukherjee—requested Prafulla Chandra to take up the responsibility of setting up the postgraduate Chemistry Department, to which the ever-enthusiastic savant agreed. In 1916, aged 55, he retired from his position at the Presidency College and joined the new Chemistry Department at the University as its Head and as the newly-founded Palit Professor of Chemistry. A new building to house these Departments came up at 92 Upper Circular Road, right next door to Prafulla Chandra's rented house, and it was here that he moved, to stay in a single room at the University College. As he grew older, Prafulla Chandra's life grew increasingly ascetic, and his substantial earnings mostly went into donations either for research or for philanthropy, until he was universally hailed as *Acharya* and described by no less a person than Gopal Krishna Gokhale as India's 'scientist-saint'. He lived into extreme old

---

[4] It was registered as a limited company in 1901. In 1980, it was nationalised as a public sector unit and renamed the Bengal Chemicals & Pharmaceuticals Ltd. (BPCL).

age, dying at the age of 83 in the same room in the University building where he had lived for three decades.

Always a patriot and a fighter for freedom, Prafulla Chandra travelled widely, lecturing and encouraging young Indians to stand up for themselves and equal the ruling Europeans in ability as the best way to claim Independence. Perhaps his dream is best expressed in his own words to the Indian Science Congress of 1920, where he presided. The *Acharya* said *"...if India, by the grace of God, will avail herself of this opportunity to rise equal to the occasion, if her men of Science and industrial pioneers will put their shoulders to the wheel together, if the study of Physics and Chemistry, of Mining and Engineering, of marine and Aerial Navigation and of the Biological Sciences will succeed in enlisting on their behalf the energy and enthusiasm of thousands of votaries, if the young men of the middle classes will crowd in great numbers the science colleges and the technological institutes more than the law colleges, if the scientific services of the State be thoroughly Indianised, if her rich men will award more scientific scholarships and establish technical schools, India will not take a long time in coming to the forefront of nations and making her political renaissance not a dream but reality."* This is a true measure of how well the venerable scientist understood the people whom he loved so well.

4.2 The Genius of Jagadis Chundra Bose

Our spotlight next shifts to the other *Acharya*, Jagadis Chundra Bose, who was to physics and biophysics in India what Prafulla Chandra Ray was to chemistry and pharmaceuticals. In the early decades of the twentieth century, his fame had spread all over the world, and it seemed that, just as Rabindranath Tagore would bring the ancient wisdom of India to the West in the sphere of letters, so would Jagadis Chundra Bose in the field of science. Soon, however, in the chauvinistic 1930s and the war-torn 1940s, the reputation of these Indian sages underwent an eclipse, and they were either ignored or their ideas treated with disdain. This lasted for decades. It was only in the last part of the twentieth century that their reputations partially emerged from these shadows, when the end of the Cold War brought forth a wave of generosity in the western world,[5] and some belated credit began to be given. We live again in less generous times, but it is fitting that there should be a sane judgment and a more objective assessment of the achievements (and failings) of the man who was arguably the greatest scientist produced by India since the days of Aryabhatta.

Scion of a landowning family from Munshiganj (ancient Vikrampur) in the Dhaka district of East Bengal, Jagadis Chundra Bose was born in 1858 to a wealthy deputy magistrate under the British crown, and his wife. Educated initially at Faridpur and Calcutta, he travelled to London in 1879 to study medicine. Unable to continue because of recurring bouts of *kala azar* (leishmaniasis), an ailment picked up on

---

[5] This roller coaster of reputation, and its socio-political origins, has been nicely analysed by Amartya Sen in his essay on Tagore in *The Argumentative Indian* (Penguin Books, 2006).

a youthful hunting expedition, he turned to pure science, winning a scholarship to Cambridge. Under the tutelage of the venerated Lord Rayleigh, he completed his studies, winning a D.Sc. in 1896, prior to his return home. On Rayleigh's recommendation, Lord Ripon, Viceroy of India, appointed him to a junior professorship at the Presidency College in 1895. Here he remained till the establishment of his own Institute in 1917, the same year as he was conferred a knighthood by the British Government. He remained Director there till his death in 1937, at the age of 78.

Bose's real claim to fame lies, like that of many outstanding scientists of his generation, in his amazing abilities as an inventor and instrumentalist, often using the most mundane objects to produce startling results. He lived in an age when electricity and electromagnetic phenomena were considered to be the key to the secrets of the Universe. The bulk of Bose's oeuvre was concerned with making extremely sensitive measurements of minute quantities of electricity. His research may be classified under two distinct headings—his investigations of electromagnetic waves, and his study of plant and metal responses to external (especially electric) stimuli. The first class of work has been fully accepted by West and East, and lies enshrined in the IEEE's 2012 milestone declaration of Bose as a father of radio technology. The second class of work—violently contested, purloined, debunked and ignored in succession over the past century—was certainly what made Bose a household name in the first half of the twentieth century and an intellectual pariah in the second. Over the past three decades, however, many of his empirical results have been shown to be correct, and some of his more conservative conclusions have received a grudging acceptance. Experts continue to greet the more radical ones with a derisive smile, and semi-experts to echo that skepticism without a second thought. Let us, then, see for ourselves what this great man did and thought, and draw our own conclusions.

The existence of electromagnetic waves had been proven by Heinrich Hertz's discovery of invisible 120 nm radiation at Berlin in 1886. Hertz died young in 1892, leaving the mantle to his successors, Sir Oliver Lodge in Birmingham, Nikola Tesla at New York, Augusto Righi and his student Guglielmo Marconi at Bologna, Ferdinand Braun at Strasbourg, Alexander Popov in St. Petersburg—and Jagadis Chundra Bose in Calcutta. The fact is that all these brilliant men were able to generate Hertzian (or should it be Maxwellian?) waves of different wavelengths and use them for some sort of communication. The debate over priority still rages, but the timeline is more-or-less as follows: Hertz (1886), Lodge (1888), Tesla (1891), Righi (somewhere between 1890 and 1894), Bose (November 1894), Marconi (December 1894), Popov (1895), Braun (1897). It was, apparently, Lodge's 1894 memorial lecture to the Royal Society following Hertz's death, where he demonstrated both the generation and detection of Hertzian waves, which spurred many of the younger inventors to repeat this work on their own. There is no question that it was Marconi—a man with no scruples about patenting the ideas of other inventors—who pursued the practical applications and converted these scientific curiosities into the engineering marvel which has come to stay as part of our everyday life. The Nobel Committee, which honoured him with its prize in 1909, perhaps recognised his go-getting abilities above his scientific ingenuity. However, the fact that Marconi's prize was shared with Braun, for inventing the first semiconductor diode rectifier, is where the shoe pinches, for it is now well-established that Bose was the first to do this—and he did not hide his light under a bushel. That Braun could have won the Nobel Prize anyway for his invention of the cathode ray tube (CRT) is quite beside the point.

What Bose was really able to do was to generate what we would today call microwave radiation at a wavelength of around 5 millimetres, whereas Lodge, Marconi and others were using wavelengths around 100 metres or more. The penetrative power of these long wavelengths was much greater, and this enabled Marconi to use them for long distance signalling. In particular, Marconi's waves could go all around the world, for reasons which he did not understand at the time. However, Bose's waves were much more conducive to studying their light-like properties, which was his main interest. While proving that these waves could undergo the usual phenomena of reflection, refraction, polarisation, etc., Bose invented some of the commonplaces of modern radio technology, including waveguides, the horn antenna, dielectric lenses and polarisers, and, most famously, improved radio wave detectors. He displayed remarkable ingenuity in using the simplest of materials for his researches—for example, a dielectric polariser was constructed by taking a fat Bradshaw railway timetable and interleaving its pages with tin foil.

It was Oliver Lodge who had perfected and demonstrated the first iron 'coherer', which we would now call a radio wave detector. This improved greatly on Hertz's primitive spark gap, but it had its own problems. It basically consisted of an evacuated glass tube filled with iron filings, with two electrodes at the two ends. When electromagnetic waves passed through it, the iron filings tended to stick together (a discovery made earlier by the French physicist Edouard Branly), and lower the resistance of the device, which would then cause a current pulse in a closed circuit with a galvanometer whose needle would show a kick. However, after this the iron filings would remain stuck together, and would have to be recovered or 'decohered' by tapping the device with a clapper. This worked fine if the electromagnetic signal came as short pulses separated by relatively long gaps, but when they came thick and fast, all that would happen was that the clapper kept quivering and the galvanometer needle stayed deflected. Bose's invention— demonstrated before the Royal Society in 1899—was the so-called 'iron-mercury self-recovering coherer', which consisted of an evacuated U-tube of mercury with an iron disc closing one end and a plunger the other. The plunger was slowly worked in with the help of a screw arrangement, and the mercury meniscus pushed into delicate contact with the iron disc. This worked fine as a detector of rapid-fire pulses, for the resistance fell and jumped back faithfully as the electromagnetic pulses arrived and went. The sensitivity could be adjusted by pushing or releasing the plunger, bringing more or less of the mercury in contact with the iron disc. The current pulse was transduced into an audible click by using a telephone receiver, whose carbon granules had a much faster recovery time than a galvanometer circuit. It is now believed that the real cohering action and its quick recovery happens in a layer of oxide at the point of contact between the two metals. Bose guessed that this was 'due to a rearrangement of the surface molecules' at the interface, though the oxide part eluded him (and his contemporaries), and his 1899 paper was mainly an account of the various metal pairs he had tried before he hit upon iron and mercury as the most convenient pair. There can be no doubt that Bose's radio-wave detector was a significant improvement on anything invented before.

In a letter to his friend and fellow-intellectual, Rabindranath Tagore, back in Calcutta, Bose wrote that the millionaire proprietor of a telegraph company had come to him, patent form in hand, just before Bose read his paper to the Royal Society and begged him not to reveal all the details to the audience. *'There is money in it'*, he had pleaded, using his strongest argument. The magnate was probably totally sincere, but what he did not realise was that this approach only served to alienate the high-minded Indian inventor. Bose not only refused point-blank, but grandly declared that his invention was its own reward and others were free to use it. He then went back to India, head held high. Two years later, when Marconi startled the world by catching in England a radio signal sent from across the Atlantic, his detector was a mercury-iron coherer *à la* Bose. The only difference was that this was a straight tube with a globule of mercury in it, which was pushed by an air column to touch an iron plate. Marconi then applied for a British patent for this *in his own name*. Maybe Bose had not reckoned with the fact that Marconi would take his magnanimous offer so literally. It must be noted that Marconi was entirely within the letter of the law in patenting *his* version, since Bose had not patented his. The silence of the British scientists, who knew perfectly well who was the originator of the mercury-iron coherer and the telephone detector, was deafening.

The hue and cry over Bose's right to Marconi's prize has arisen only a century after the fact. Bose himself did not raise this issue anywhere. By then he was involved in a new area of research altogether. But before we tell that story, we must mention that he continued to work on improved coherers for some time. In the course of this, he discovered that a crystal of galena could act as a coherer because of what we today call its semiconducting properties. In 1909, Marconi's Nobel Prize was shared by Braun, essentially for inventing the 'cat's whisker coherer', a semiconductor diode made of lead sulphide. The common name for this mineral is—galena! It is not so well known that Bose, prodded by his benefactor Mrs. Bull, had actually filed for and won an American patent on this galena coherer in 1901—showing that he was not averse to patenting when there was no strident millionaire to irritate him. Bose, in fact, had gone further and experimented with selenium, another semiconductor, and even had some inkling of the effect of p and n type impurities. However, it would be improper to ascribe the semiconductor revolution to him, since the key feature of that is the *p-n junction*, something that came up only in the 1940s. Nevertheless, in many ways, Bose was, as the Nobel Laureate Sir Nevil Mott, said later, *'some 60 years ahead of his time'*.

One of the discoveries of Bose was that his metal-contact coherers would lose their sensitivity if used continuously for some time, but would recover if left alone for some time. The analogy with fatigue and rest in a living being was very close. The phenomenon of metal fatigue was well known by then to civil engineers and was already known to have caused some disasters. But metal fatigue is a mechanical effect due to the development of microscopic cracks which grow under repeated stress and there is no question of recovery. Whereas there was both fatigue and recovery in the *electrical* properties. The temptation to make a connection proved irresistible. We must remember that electricity was the great mystery of the nineteenth century and thought to be intimately connected with life, as indeed it is, though not in the mystic way it was then thought of. In fact, the electrochemical nature of nervous action was not understood, and there was a great deal of mystery and a lot of nonsense written concerning electricity and 'the ether'. In the closing years of the nineteenth century, no less a scientist than Oliver Lodge and the hugely-popular writer Arthur Conan Doyle believed that one could communicate with ghosts by electricity. These would all influence Bose's mind, but the strongest motivation came from his Indian roots, and from his mystic friends Tagore, Vivekananda—and Margaret Noble, Sister Nivedita. We shall not pursue these influences further, but concentrate on Bose's empirical work.

From metals, Bose turned to plants and animals, and especially to plants. Plants stand still in one place and are easy to experiment with, unlike animals which would have to be tied down and tortured—something which the soft-hearted Bose was not willing to do. With his uncanny knack of being able to measure extremely small electrical signals, Bose was able to do for plants what the ECG and EEG do for human beings—measure the electrical impulses corresponding to basic life processes and especially their response to external stimuli. Among his important discoveries was the fact that sap rises in plants not merely by capillary action, but by a slow peristaltic action which he likened to the beating of an animal heart. Plants gave off 'distress' signals when tortured (by having a portion torn off or cut) or poisoned (typically with chloroform). They were found to 'remember' things like excessive cold or heat and plants like the parasitic creeper Dodder, could even 'take decisions' on which object to entwine. Plants could communicate with other plants by chemical signals. They possessed internal electrochemical pathways which were like a primitive nervous system. Above all, they could and did move. The invention that Bose was most proud of was his 'crescograph' or growth meter, which could magnify tiny movements of plants 10,000 times by a delicate system of levers. Viewed through such a machine, said Bose, '*the motion of a snail would appear like that of an express train.*' Using this novel instrument, Bose watched plants growing, saw them shudder when touched, cut or frozen, and demonstrated the 'death throes' of a poisoned plant. It was an empirical *tour de force*, a glimpse into a hitherto-unknown world, akin to Galileo's telescope or Hooke's microscope.

It should be noted here that the oft-repeated statement that Bose proved that plants have life is just not true. Even primitive folk know that plants are born (germinate) and that they die. What Bose found was that plants have minute responses to even very small external stimuli, which are very similar to animal responses. Perhaps his findings may be better stated that animal—and human—responses are closer to plant responses than outward appearances would make us think. All this was discovered by the empiricist in Bose. The poet in him led to the conclusion that plants are sentient beings, i.e. they feel pain, remember things, and can communicate and even respond to affection. And his mystic streak led him to see similarities in the behaviour not just of plants and animals, but of metals and inanimate things, eventually broadening out into an eclectic pantheism very much like the *Vedanta* of his Hindu roots. Thus, where Vivekananda arrived by meditation and Tagore by his overwhelming love for Nature, Bose arrived with his needles and smoked glass plates.

Unlike Jagadis Chundra's radio work, which was very well appreciated in the West, his work on plants met with opposition from the very beginning. Physicists appreciated the sensitive instrumentation, but plant physiologists opposed his findings violently. They attributed Bose's conclusions to his Hindu pantheism, to which they opposed the impervious shield of their Protestant dogmatism. With hindsight, we can now see that while some of their criticism originated in Christian orthodoxy and not a little in racial envy that a man from a colonised country could hold his own among his 'betters', the core dispute arose from sheer incredulity at the novelty of Bose's results. If his work on semiconductors was sixty years ahead of his time, surely his work on plant physiology was a hundred years ahead.

While Bose's work was given scant attention in England, this was nothing compared to the difficulties and harassment he faced at home, at the Presidency College. Much of his research had to be financed from his own pocket, which perhaps explains his use of cheap household materials. When he applied for a research grant, a sneering administrator wrote that '*a native who cannot make do with 500 rupees a month must be having his head a bit turned*'. The climax came when Lord Rayleigh, then President of the Royal Society, came to Calcutta, and Bose took him to see his laboratory in Presidency College, where his Lordship spent the whole day in scientific discourse. The next day, Bose found himself served with a show-cause notice from the college Principal, asking why he had permitted 'an outsider' to enter a College laboratory, without seeking permission from his own august self.

This brings us to the final item in this story. When Swami Vivekananda lectured in America in 1893–94, he acquired, among others, a devoted follower in Mrs. Sarah Chapman Bull, a wealthy widow with a great interest in India. In the following year, he lectured in London, and acquired another fan—Miss Margaret Elizabeth Noble, the future Sister Nivedita. This trio visited Bose's laboratory in 1898 and watched him demonstrate his inventions at Paris in 1900—and were suitably impressed. A close friendship sprang up between the two ladies and the Boses, which lasted till both *memsahib*s died in 1911. Egged on by Sister Nivedita, Mrs. Bull, whom the scientist would always address affectionately as 'Mother', granted him the then-enormous sum of 20,000 dollars to aid his scientific research. When he wrote that most of this sum had gone in purchasing land for the new institute, she gave him another 20,000 dollars. It was this money that built the Bose Institute—India's first research institute dedicated to pure science, if one does not count the Indian Association. At this juncture, the British Crown knighted Bose and the Royal Society, after having ignored his work for two decades, elected him a Fellow. Alas! Bose was now more than 60 years old and his inventive years lay well in the past.

Eventually, like *Acharya* Ray, *Acharya* Bose's greatest contribution lay in his legacy. He inspired a generation of brilliant students like Satyendra Nath Bose, Meghnad Saha, Sisir Kumar Mitra, Bidhu Bhushan Roy and his own nephew Debendra Mohan Bose to take Indian science to heights never before imagined. If Radhanath Sickdher had laid the foundation stone, the two *Acharya*s built the temple of Indian science brick by brick, and their brilliant successors left it clothed in marble. After Independence, we have only been able to embellish it with a carving or two.

4.3 Postgraduate Science

From its inception in 1857, the University of Calcutta had a series of distinguished and not-so-distinguished Vice Chancellors, mostly jurists and civil servants. In 1906, amidst the tumult of the anti-partition agitation, the position was given to an Indian jurist, who was also a mathematician of note. His name was Asutosh Mookerjee, later Sir Asutosh.

Sir Asutosh (1864–1924) came of a family of Brahmin scholars. He was descended from *Pandit* Ramchandra Tarkalankar, a noted Sanskrit scholar, who had held the first chair of *Nyaya* philosophy at the Sanskrit College. His own father, Ganga Prasad, was a flourishing doctor, and an educationist who founded the South Suburban School in Calcutta. Brought up in an atmosphere of erudition, the young Asutosh showed an early aptitude for mathematics. At the age of 15, he passed the entrance examination of the University of Calcutta, standing second and receiving a scholarship. He was admitted to the Presidency College, where he took mathematics as his subject. At the tender age of 16, he published his first paper—a new proof of a theorem of Euclid. It is probably the first paper to be published by an Indian in a western peer-reviewed journal. In 1885, the young prodigy completed his graduation in mathematics, and then, a year later, he took a second degree in Natural Sciences, making him the first Indian ever to get a dual degree from a University. In the same year, his third publication, '*A Note on Elliptic Functions*', proving one of the several 'addition theorems' for this important class of special functions, earned him high praise from none other than Arthur Cayley, the famous mathematician who was then the President of the British Association for the Advancement of Science.

Back in Calcutta, Dr. Mahendra Lal Sarkar spotted the talented youth, and employed him as a lecturer at the Indian Association. Failing to get a university chair, however, the disappointed young mathematician turned to the study of the law, then one of the most lucrative professions open to Indians under the British dispensation. After 1893, therefore, Asutosh Mukherjee was lost to mathematics, but he became an ornament of the courts, an upright judge, and finally, an iconic Vice Chancellor of the University which could not give him a teaching position.

Once he became Vice Chancellor, Asutosh was now free to determine the future of his University, but had not forgotten his early days as a struggling mathematician. He determined, therefore, that the University would raise itself up, under his guiding hand, from an examining and degree-granting body, to a genuine temple of learning, like the Universities in Britain and the West. For this, it was necessary to found postgraduate departments, where higher studies and research would be carried out, so that the University could award Master's and doctoral degrees. The Viceroy and the Government of Bengal were sympathetic, but—an oft-repeated theme—they were not willing to commit money beyond a token sum.

And so, like many an academic leader before and after, Asutosh became a vigorous fund raiser. Money poured in, from opulent landlords like the *Maharajas* of Darbhanga and Cossimbazar, rich lawyers like Joykissen Mookerjee, business magnates like *Seth* Premchand Roychand and from the general public, and soon Asutosh found himself sufficiently solvent to start the postgraduate Departments he had dreamed of. The biggest contributors were two fellow lawyers. Sir Rashbehari Ghose, who had made a huge fortune in the courts, had also held the Tagore Chair of Law at the University during 1875–76. He eventually donated a princely sum of 2.1 million rupees, the equivalent of about 33 million US dollars at today's rates of exchange. Sir Taraknath Palit, comparatively less wealthy, but even more dedicated, poured his entire fortune of 1.4 million rupees into the University (equivalent to 22 million dollars), reducing himself to genteel poverty as *Rai* Kedar Nath was to do in Delhi a few years later. These contributions dwarfed the royal contribution of 250,000 rupees which the University received from *Maharaja* Rameshwar Singh of Darbhanga.

With such funds in hand, Sir Asutosh went full steam ahead to found postgraduate departments in the University for a wide variety of subjects, including Physics, Chemistry 'Mixed Mathematics' (later to become Applied Mathematics) as well as innovative subjects like Comparative Literature, Applied Psychology, Industrial Chemistry, Ancient Indian History and Culture and Islamic Culture. Post-graduate courses and research were started in English, Sanskrit, Pali, Arabic, Persian, 'Mental & Moral Philosophy', History, Economics and Mathematics. A brand new University College of Science, built in the by-then ubiquitous Indo-Saracenic style developed by the colonial architects, arose where Sir Taraknath used to reside, for he had donated his own dwelling house to the University.

Armed with all the money and support, Sir Ashutosh, as he became after 1911, built up a whole series of truly stellar departments in his new model University. No institution in India before or after has ever had such a galaxy of talents. Our focus, however, is on the science departments, and, in fact, on the Department of Physics, where the first glimmerings of particle physics research began to show up. The description of these halcyon days of the University merit a whole volume in itself, but a very brief sketch of the leading lights in Physics is attempted in the next section.

As for the great Vice Chancellor, whose dominating personality and powerful voice earned him the epithet 'Tiger of Bengal', he served the University in four successive two-year terms from 1906 to 1914. After a seven-year hiatus, he was again appointed Vice Chancellor in 1921. Refusing to serve a sixth term, as he could not see eye to eye with the Governor of Bengal, Lord Lytton (son of a Viceroy and grandson of a famous author), the doughty educator quit his post and also the judicial service, and took up his advocate's gown and wig again. Once again his great voice could be heard, dominating the courts. The 'Tiger' died suddenly—perhaps as he would have liked—in harness, while arguing a case before the Patna High Court. He missed his sixtieth birthday by a month.

4.4   The World Beaters

For the new postgraduate Department of Chemistry, Sir Asutosh had a leader ready at hand—none other than the revered *Acharya* Prafulla Chandra Ray. The other *Acharya*, however, already had his own Institute, and so, for Physics, the farsighted Vice Chancellor decided to go for youth. His first appointment to the newly-minted Ghose Professorship was the nephew of the great J.C. Bose, Debendra Mohan Bose, or D.M. Bose as he was widely known. However, D.M. Bose went away to Germany soon after his appointment in 1914, and did not return till 1919. In the meanwhile, a group of talented youths were appointed as Lecturers in the new post-graduate Department. Their names are legendary—Satyendra Nath Bose, Meghnad Saha, Sisir Kumar Mitra, Bidhu Bhushan Ray, and Sailen Ghosh. It is related that the Vice Chancellor called Ghosh, Saha and Bose to his office and asked them what they could teach. To which they chorused *"Whatever you say, sir!"* And so they were tasked with teaching the very latest developments in physics—relativity, quantum theory, crystallography, and so on. Prasanta Chandra Mahalanobis, who would later go on to found the Indian Statistical Institute, and was then already teaching at the Presidency College, was roped in to teach the postgraduate classes. In 1918, he was already teaching an elective course in the General Theory of Relativity, proposed by Einstein in 1915. It is rare that a Masters' programme contains subjects so much in tune with contemporary research.

In 1917, Sir Asutosh made an unusual appointment. He persuaded an officer of the Audit and Accounts Service, posted at Calcutta, to quit his job and join the University as Palit Professor of Physics. This officer, who was doing research part-time as an evening hobby at the Vizianagaram Laboratory, was none other than Chandrasekhara Venkata Raman, or C.V. Raman, the future Nobel Laureate. Even though this means that his monthly salary would drop from 1,100 rupees to 600 rupees (for that was all that could be given from the Palit endowment), the young Raman jumped at the opportunity. Events proved his decision to be the correct one. However, it would be 1929 before he and his student K.S. Krishnan discovered the famous light-scattering effect which bears his name. The Nobel Prize followed in 1930. In 1933, Raman left Calcutta University forever. He had been appointed Director of the Indian Institute of Science in Bangalore—the first Indian (and the only Nobel Laureate) to hold the position.

Sailendranath 'Sailen' Ghosh, who had been a brilliant student (standing first in Physics, as S.N. Bose did in Mixed Mathematics), was deeply involved in the freedom movement and had active links with the terrorist groups *Anushilan Samity* and *Jugantar*. He was entrusted with setting up teaching laboratories in the new Department. However, his tenure at the University was cut short when he learnt that the police were looking for him and that a grim future at the dreadful penal colony in the Andaman Islands awaited him. After some hair-raising adventures, he ended up in America and became a full-time revolutionary, spending the next twenty years of his life mobilising public opinion against the British *raj* and its excesses.

Meghnad Saha was the first of these young Turks to make an international name. This son of a village grocer had overcome the twin barriers of his family's poverty and low caste to become one of the brightest sparks in the University and a protégé of the formidable Vice Chancellor. In those days, his interests lay in thermal physics, statistical mechanics and quantum theory. With his colleague, S.N. Bose, he had come up with an equation of state for a non-ideal gas (1918). But this was a mere preliminary. In 1920, the *Philosophical Magazine* published his paper on *Ionisation in the Solar Chromosphere*, where he set up an equation which relates the intensity of spectral lines of a star with its temperature. The Saha Ionisation formula, as it is now known, quickly became a staple of astrophysics and is now used for superhot plasmas in general, including the early Universe. It made its young discoverer a celebrity overnight. Awarded a Premchand Roychand scholarship, Saha sailed to England, to reside for some time at Cambridge and explore some of the far-reaching consequences of his discovery. Returning to Calcutta, he did not stay long, but decided to leave for Allahabad University in 1923.

Satyendra Nath Bose, who had stood first in Mixed Mathematics, was another prize catch, and he devoted himself to teaching relativity, then a new and (always!) exciting subject. With his former classmate and now colleague, Meghnad Saha, he published the first English translation of Einstein and Minkowski's pioneering papers in 1920. Bose relates that they had no access to modern textbooks till they found that P. J. Bruhl, a botanist-turned-physicist at the Bengal Enginering College, had a very up-to-date collection of physics books, which he was only too glad to lend to these young enthusiasts. Reading Max Planck's works required learning German, which the duo obviously put to good use. But Bose, like his friend, did not stay in Calcutta University for very long. In the words of D.M. Bose, *"During 1920s the situation in Physics Department of the Science College was becoming rather uncomfortable; there were too many able scientists crowding together who were provided with inadequate laboratory accommodations, technical resources and apparatus; consequently a certain amount of heat was generated... The first physicist to migrate was Satyendra Nath Bose."*

To Dacca (now Dhaka, capital of Bangladesh) went the young physicist, to join the Dacca University which was being founded by a dynamic Vice Chancellor, Sir Philip Hartog, who rivalled Sir Asutosh in vision and ability to pick talent. The 'Tiger' had now quit his position at the University, and Hartog (and others) felt free to poach faculty from the already disintegrating Departments. It was at Dacca in 1923 that Bose made his famous derivation of Planck's formula for blackbody radiation, in which he used only statistical arguments and no specific interactions between photons and matter. It immediately trumped work being done by stalwarts like Ehrenfest, Einstein, Pauli and Sommerfeld, but was rejected by the *Philosophical Magazine* for unknown reasons. Bose then sent his manuscript to Einstein, who at once recognised its deep insights, and translated it himself for publication in a German journal. Einstein followed up Bose's ideas to set up what are now called the Bose–Einstein statistics, which are the foundation of quantum statistical mechanics. This was perhaps the only incident in the career of the great man where he extended a brilliant idea due to someone else. Having his name coupled with Einstein, of course, made Bose an all-time celebrity, and when Paul Dirac coined the name *boson* to indicate particles (like photons) that obeyed his statistical formula, it put his name into practically every modern textbook.

Slightly older that these two stalwarts was Sisir Kumar Mitra, who had been leading a life of penury and frustration as a college lecturer in a small district. It was again Sir Asutosh who had been impressed by some of the popular science articles written in the vernacular by the young teacher, who brought him to the University to complete his doctoral studies under Raman. This gave an opportunity for his talents to bloom, and thus Mitra was awarded a D.Sc. in 1919. He then went to Europe, completed a second doctoral degree under Fabry, worked briefly with Mme. Curie and then joined the laboratory of Camille Gutton to study the physics of radio waves and their detection. This made him an expert in electronic valves and their applications. On his return to Calcutta, he persuaded the encouraging Sir Asutosh to let him open a new department which eventually became the Department of Radiophysics and Electronics. Unlike Bose and Saha, who returned to Calcutta University only after spending many years elsewhere, Mitra devoted his life to the same University. His discovery of the 'D-layer' in the ionosphere earned him international fame—enough to have a lunar crater named 'Mitra' in his honour.

Bidhu Bhushan Ray, or B.B. Ray as he is better known, was a classmate of the ebullient Sailen Ghosh, but devoted himself exclusively to science. He was also a student of the versatile Raman, and his doctoral work on unusual optical phenomena in the atmosphere, such as coronas, glories and noctilucent clouds is still relevant. After obtaining his D.Sc. in 1922, Ray moved to Europe under a Premchand Roychand fellowship, working with Nobel Laureates like Manne Siegbahn at Uppsala, Niels Bohr at Copenhagen and Gerhard Herzberg at Darmstadt. On his return to Calcutta, he set up a labratory for X-ray crystallography, the first of its kind in India. Among the bright sparks who started their training in his laboratory was R.C. Majumdar, who went on to become a particle theorist of note. Ray's most significant discovery was the Raman effect in X-ray scattering, but his later days were darkened by a controversy with American spectroscopist J. M. Cork, who denied that such an effect existed. Though vindicated by a magisterial analysis by Arnold Sommerfeld, Ray may have taken the controversy to heart, for he died suddenly in 1944, aged just fifty. Over the years, he has largely been forgotten. In his own University, where Ray spent his life, all that exists is a portrait in oils hanging in a lecture theatre, and a mnemonic handed down from one generation of students to the next—"*B.B. Roy of Great Britain had a Very Good Wife*"—which stands for the code by which we can read the value of a resistor from the coloured bands around it. One can sense some light badinage in this mnemonic, but unfortunately no one living today knows what it was all about. *Sic transit gloria mundi.*

With this young brigade of scientists, all contributing at the highest level, one may say that the movement begun by David Hare and Ram Mohun Roy had reached its consummation. From eager acolytes at the temple of Western science, Indians were now setting the score, making discoveries to match those of Western scientists and etching their names into science textbooks. This section has talked only of physicists, but similar developments occurred in Chemistry, Botany and most of all Mathematics, where the name of Srinivasa Ramanujan will remain in letters of gold forever.

Western science in India had truly come of age.

Part II
Particle Physics—The Early Days

Chapter 5
The Beginnings

5.1 Elementary Particles

The physics of elementary particles, which is nowadays called High Energy Physics, is, in a sense, as old as the atomic theory, i.e. it goes back to the philosophers Leucippus and Democritus in Greece and to the sage Kanāda in India. For the quest has always been the same: to find the elementary constituents of matter, which are believed to be discrete, and to find out how, by interacting between themselves, these objects can conglomerate to form the substances which, in turn, make up the material world. Till the seventeenth century, however, it remained a metaphysical quest and was not of much interest to practical scientists. It was the great Isaac Newton, an accomplished chemist, or perhaps alchemist, who, in his precise way, began to make meticulous measurements of the weights of different chemicals which went into a reaction, expecting to extract from these numbers something of the laws that he instinctively knew underlay their behaviour. It was not given, however, to the same man to be the father of Chemistry as well as of Physics, and thus Newton's chemical (or alchemical) researches proved as barren of results as his researches in Mathematics and Physics were fruitful. All it brought him was recurring bouts of madness which came from inhaling too much mercury vapour, which is known to cause personality disorders.[1]

---

[1] Hallucinations and personality changes due to the inhalation of mercury vapour are also known as 'mad hatter's disease', since it was common among hat makers, who then employed elemental mercury to cure the felt used to line the insides of hats. A good deal of mercury vapour was inhaled in the process, and its poor victims gave rise to the expression 'mad as a hatter'. Hats have gone out of fashion, and so most readers know only the Mad Hatter in Lewis Carroll's *Alice in Wonderland*.

© The Author(s), under exclusive license to Springer Nature Switzerland AG 2021
S. Raychaudhuri, *The Roots and Development of Particle Physics in India*,
SpringerBriefs in History of Science and Technology,
https://doi.org/10.1007/978-3-030-80306-3_5

Newton's hunch was proved correct, however, like many of his other scientific intuitions. Over the next century, four laws of *stoichiometry* were discovered, enshrining precisely the rules by which fixed amounts of elementary substances (for by then they had been defined as such by Robert Boyle, Newton's senior by 15 years) came into combination to form compounds, as they later came to be called. Their discoverers—Antoine-Laurent de Lavoisier, Joseph-Louis Gay Lussac, Joseph-Louis Proust and Jeremias Benjamin Richter—who all came from continental Europe, succeeded where the English genius had failed. Fortunately, Newton did not live to see this, for, in his own words, he did not *'love to be dunned and teezed* (sic) *by foreigners about mathematickal* (sic) *things'*. To balance it out, it was an Englishman, John Dalton, who made sense out of these empirical laws by showing that they indeed point to a granularity of matter akin to the 'atoms' of Democritus. And thus were modern atoms born. They remain the elementary particles of all chemistry.

Another century passed after Dalton, and the discovery of radioactivity put a new complexion on elementarity. Sir J.J. Thomson had just discovered the electron, which seemed far smaller than the lightest atom, and also seemed to be a constituent of all atoms, irrespective of other properties. Now radioactive atoms seemed to disintegrate, emitting electrons and also particles which turned into helium nuclei, so it became clear that atoms are composite objects. Rutherford, who discovered this, was said to have 'split the atom'. Soon he went on to make an even more momentous discovery—that of the nucleus. Combining this with an inspired guess made earlier by the Rev. William Prout, he declared the hydrogen nucleus to be an elementary particle as well—the proton. The discovery of the neutron by Chadwick in 1932 completed the triad of elementary particles which concern atomic and nuclear physics—the electron, the proton and the neutron.

The march of granularity did not end here, however. A succession of new discoveries followed, of more particles, and eventually another layer of compositeness for most of them. In modern parlance, therefore, elementary particles generally mean particles which are 'smaller' than the atomic nucleus and often have such fleeting lifetimes that they can be produced and detected only in highly sophisticated experiments. So it is not surprising that modern particle physics grew as an offshoot of nuclear physics and radioactivity studies in the first half of the twentieth century and that it was only by the middle of the century that it had begun to assume a separate character. Our story begins, therefore, from these 'radioactive' times.

5.2   Radioactivity in Bombay and Bangalore

Radioactivity was discovered, quite by accident, by Henri Becquerel in 1896 and later studied by the Curies, Pierre and Marie. This achieved wide publicity when the Paris trio shared the 1903 Nobel Prize in Physics. At the time, radioactivity was thought to be some sort of fluorescence phenomenon and hence it was considered more a branch of optics. It was the path-breaking experiments of Ernest Rutherford and Frederick Soddy, carried out at McGill University in Canada between 1898 and 1902, which established it as a sub-atomic phenomenon, and led to Chemistry Nobel Prizes for Rutherford in 1908, Mme. Curie in 1911 and Soddy in 1921. The existence of the atomic nucleus was announced by Rutherford in a 1911 paper in the *Philosophical Magazine*, where he published his famous scattering formula and showed how it fitted the alpha particle scattering data painstakingly gathered by Geiger and Marsden between 1909 and 1911. This was a theoretical *tour de force* by Rutherford, the quintessential experimentalist who never made a secret of his contempt for 'mathematicians', as he liked to describe theoretical physicists.

These developments did not go unnoticed in India. The first person to enter the field was a Jesuit priest at the St. Xavier's College in Bombay, Fr. Adolphus Steichen, a Göttingen graduate, who was then teaching physics at the iconic south Bombay college. With his colleague Fr. H. Sierp, the professor of chemistry (another Göttingen product) he conducted a survey of radioactivity at the hot springs at Tuwa, north of Bombay, in 1910, and they published a paper on this in the *Philosophical Magazine* in 1916. The main purpose of this paper was to contradict a hypothesis put forward by the American professor R.R. Ramsey of Indiana University, who had claimed that the radioactivity of hot springs (there are plenty in Indiana) increases with the seasonal flow of water and decreases when the flow diminishes. The paper by Steichen and Sierp reported just the opposite at Tuwa. The duo correctly interpreted the radioactivity as due to water erosion of some seam of radioactive mineral in the rocks washed by the stream. At Tuwa, the erosion rate was more or less constant, and thus increased water flow caused dilution. In Indiana, claimed the paper, increased flow washed out more of the radioactive material, leading to Ramsey's results. Subsequent studies have proved the Jesuit fathers to be largely correct.

An interesting aside is the fact that this work was done in 1910, and the paper took six years to be published. Part of this is because the authors were busy with their teaching work and could only prepare their manuscript in 1914. But just then the First World War broke out, and being German citizens, the two priests became 'enemy aliens' in British-controlled India. They were interned in a camp, and not permitted to send out their paper till 1916, when it promptly got accepted.

From the point of view of publication, however, Father Steichen had been pre-empted by a chemist, Herbert E. Watson, from the new Indian Institute of Science (IISc) at Bangalore (now Bengaluru). Watson and his assistants Gosta Behari Pal and W.F.S. Smeeth (of the Geological Survey of India) conducted a series of tests in 1914 of the rocks at the nearby Kolar gold fields to see if they exhibited significant signs of radioactivity, and published their results in the same year, first in the new *Journal of the Indian Institute of Sciences* and then in the not-so new *Philosophical Magazine*. They noted that there was no sign of radioactivity in the rocks at the depth to which they had gone. Watson later left the Indian Institute of Science in a huff when Sir C.V. Raman was appointed Director (1933). Probably he felt superseded, for he made a disproportionately large issue out of funds which Raman diverted from the Chemistry Department to the Physics Department, and resigned.

Fr. Steichen surfaces again as the Principal Investigator of a project to study the hot springs in the entire Madras Presidency. It was a five year project (1918–23) and the funding was exactly 1,000 rupees, which translates to about Rs.180,000 in today's currency. It was a reasonable amount, enough for the Reverend Father to hire a team to go to various places and collect the data. However, nothing remarkable came out of this report, and by the 1920s, the interest of physicists was moving away in other directions.

In fact, today we would classify these early studies as geophysics or mineralogy, but in those days academic disciplines were not so finely tuned. The direction of particle physics would now shift to a different direction altogether—to the sky.

5.3   The Discovery of Cosmic Rays

Cosmic rays were discovered, more by accident than design, by Viennese physicist Viktor F. Hess, in 1912. Hess had been appointed in 1910 as 'Assistant'—a junior scientist position—at the Institute for Radium Research in Vienna. In those days, radiation in the atmosphere, like radiation in hot springs, was a topic of great interest. It was known that Guglielmo Marconi (who had just been awarded the Nobel Prize in 1909) had been able to transmit radio signals across the Atlantic Ocean, well beyond the range allowed by the curvature of the Earth. The canny Italian had guessed that the radio waves were somehow getting reflected back from the upper atmosphere. The theories of Oliver Heaviside in England and Arthur E. Kennelly in America suggested that the reflection was due to a layer of ionised gas in the 'stratosphere', as the entire upper atmosphere was loosely called at that time (1902). This layer was named the *ionosphere* much later, in 1969. Franz Linke, a German meteorologist, was the first to verify this ionisation, by the direct but rather clumsy method of carrying electroscopes into the upper atmosphere by balloon (1903) and observing that they slowly discharged in the ionised air.

It was believed that radioactive emanations from radio-isotopes in the Earth's crust were responsible for the ionisation of the atmosphere. If so, the intensity of ionisation should decrease as one goes higher, more or less in the same way as atmospheric pressure decreases in Laplace's exponential decay law. A few observations were made by Swiss physicist Albert Gockel, in which, like Linke, he took electroscopes up in balloons (1909), but he detected no such decrease. Gockel's results were considered error-prone and not quite trustworthy, mostly because he had not accounted for the drop in atmospheric pressure at high altitudes. Going the other way, the Italian Domenico Pacini had submerged electroscopes in the water of deep lakes (1911) and found a significant *decrease* with depth, which was correctly interpreted as due to the absorption of surface-level radiation by the water.

With these results before him, Hess set himself the task of precisely verifying the level of atmospheric ionisation as the height above the Earth's surface increases. For this purpose, he used a slightly more sophisticated apparatus, in which an electroscope was placed in a chamber whose pressure was kept constant by a piston-spring device, so that the result was not affected by changes in atmospheric pressure. This type of constant-pressure electroscope had been designed by Fr. Theodor Wulf in 1909, improving on the famous Elster-Geitel design. It contained a pair of charged plates with same-sign charges, whose mutual repulsion would bend a pair of quartz fibres. As the ionisation of the gas in the chamber increased, the plates discharged faster, the repulsion decreased and the fibres straightened up faster. This effect could be observed and measured with a travelling microscope.

With this rather cumbersome apparatus, Hess made a series of balloon flights and, despite having to breathe through an oxygen mask, managed to obtain good quality data on the vertical profile of atmospheric ionisation. What he found was that after decreasing initially, the ionisation level began to rise after about 4,500 ft. This confirmed a stray result obtained earlier by Karl Bergwitz (1908). Ultraviolet radiation from the Sun seemed to be the obvious cause, since all the balloon flights took place during the day. Night flights were deemed too dangerous, since the balloonist would not know where he was descending. Hess disproved the UV ray hypothesis by making a flight to a height of 17,400 ft during the partial solar eclipse on April 7, 1912, when no diminution of the ionisation was found. He concluded, therefore, that the source must be "*a radiation of very high penetrating power, impinging onto the atmosphere from above*". In 1925, the great American experimentalist Robert Millikan named this radiation 'cosmic rays'. Hess' discovery, for which he always used the unimaginative but precise term *hôhenstrahlung* (which translates as 'altitude radiation') was honoured with the Nobel Prize in 1936.

Balloon flights in the early twentieth century were dangerous. As late as 1934, Prussian cosmic ray scientists Franz Schrenk and Hermann Masuch were found dead in the wreckage of their balloon which had drifted into Russia. It is thought that upper atmospheric currents prevented their descent and they died of hypoxia when their balloon failed to descend before their oxygen supply ran out. Nevertheless, cosmic ray research was pursued enthusiastically by scientists who were intrigued by this strange phenomenon. Hess was immediately followed at Berlin by Werner Kolhörster, who made a series of flights in 1914, verifying Hess' conclusions and carrying the data to a range as high as 30,000 ft. Though the First World War broke out immediately afterwards, the study of *hôhenstrahlung* was continued by Erich Regener, who had just got a position on the faculty of the Agricultural University of Berlin. Dr. Regener was able to design and construct a series of self-recording electroscopes, which could be sent up in unmanned balloon flights. Thus, he could collect a lot of data without endangering his assistants, and his careful work made Berlin the centre of cosmic ray research in the world. In the words of Bruno Rossi, himself a great cosmic ray scientist, "*In the late 1920s and early 1930s the technique of self-recording electroscopes... was brought to an unprecedented degree of perfection by the German physicist Erich Regener and his group. To these scientists we owe some of the most accurate measurements ever made of cosmic-ray ionisation as a function of altitude and depth.*" Thither, in 1914, came a young physicist from India, on a two-year travelling grant. His name was D.M. Bose.

5.4   The Rise of D.M. Bose

Debendra Mohan Bose (1885–1975), to give him his full name, had formidable
family credentials. His paternal uncle, Ananda Mohan Bose, was the first Indian to
become a Wrangler[2] (1874). His maternal uncle was the famous Sir Jagadis Chundra
Bose (1858–1937), who figured so importantly in a previous chapter. Trained first
by this savant-uncle at Calcutta and then by stalwarts like Lord Rayleigh, Sir J.J.
Thomson and C.T.R. Wilson at Cambridge and London, D.M. Bose came back in
India in 1913 with the current research of the times at his fingertips. Initially he found
a job at the City College, which had been established in 1881. However, Bose was
not destined to spend a long time at the City College. He was spotted by the eagle eye
of Sir Ashutosh Mookerjee, who appointed him Professor in the new postgraduate
Department of Physics (1914). He was given the Sir Rashbehary Ghose Chair of
Physics, with a lavish salary, and a generous Ghose Travelling Fellowship, which
required that he should visit a foreign country for two years and update himself on
the latest research. Bose, aged 29, immediately availed of the travel grant to go to
Erich Regener's cosmic ray laboratory in Berlin. If we recall that it was only two
years since Hess' momentous discovery, we can see that Bose was not just up to date
with the latest research, but that he also had an excellent feel for what was about to
become important.

Soon after Bose's arrival at Berlin, however, World War I broke out and, in an
inversion of the plight of the German Jesuit fathers in India, Bose, as a British subject,
was kept in loose internment till the end of the War. However, both Regener, and the
revered Max Planck stood guarantors for him, and thus he was allowed to continue his
research work uninterrupted during this period. At Berlin, Bose was able to develop a
highly sensitive cloud chamber, a device which had recently been invented by his old
mentor C.T.R. Wilson at Cambridge (1911). This was too bulky to fly in a balloon,
but he used it to photograph recoil tracks of protons produced by fast alpha particles
in the chamber. This work was published in the *Physikalische Zeitschrift* in 1916, and
was among the evidence quoted at the time to show that the momentum conservation
laws hold at sub-atomic scales. Insofar as particle physics grew out of cosmic ray
research, Bose may, therefore, be considered the first Indian particle physicist.

Stranded in Berlin for the duration of a war that seemed never-ending, Bose
made the most of his enforced exile. He started attending Max Planck's famous
lectures on classical physics, and was deeply impressed. *"After attending Planck's
lectures,* (I learnt) *what a system of physics meant in which the whole subject was
developed from a unitary standpoint and with a minimum of assumptions,"* wrote
Bose of his lessons from the maestro. Permitted eventually to return to India after
the Treaty of Versailles ended the Great War in 1919, Bose was immediately able
to resume his vacant position as Ghose Professor at the University of Calcutta. We
have already noted that it was then a department of stars, including the soon-to-be-
famous M.N. Saha and S.N. Bose. At that stage, it was D.M. Bose, rather than C.V.

---

[2] Those who passed the notoriously difficult Mathematical Tripos Examination at Cambridge with
first class honours were called Wranglers. The tradition still stands.

Raman, who acted as a mentor for these two brilliant young men and gave them rare texts out of his personal collection to read, such as Planck's *Thermodynamics*. This definitely stimulated the great discoveries which these two were fated to make.

The 1920s were a period when great discoveries were simply waiting to be made in cosmic rays, which was essentially a virgin field at the time. However, D.M. Bose was not really equipped to be a balloonist. He was the quintessential Bengali *bhadralok*, happy to work in a laboratory, but not to perform feats of derring-do in a balloon. Even at Berlin, his cloud chamber was grounded. And thus, after returning to Calcutta, Bose moved into nuclear physics and quit thinking about cosmic rays. Somehow his early enthusiasm about ionisation in the air had abated. Instead, he continued with his studies on alpha particle scattering. With his student Subodh Kumar Ghosh, he built the first cloud chamber to operate in India. This worked till 1941, when a better one was built under Bose's tutelage by M.S. Sinha and R.L. Bhattacharya at the Bose Institute.

Cosmic rays had not, however, seen the last of D.M. Bose. After two decades he was to return.

5.5 Unsung Pioneers

It was not as if the excitement of nuclear physics was confined to the University of Calcutta, even though it has figured largely in the story till now. Theoretical studies in nuclear science were initiated by an obscure teacher in the Maharaja's College at Jaipur. Named M.F. Soonawala, he was a man of wide interests in physics. Today he is perhaps best remembered as an iconic teacher at the University of Rajasthan, and for having written a magisterial description of *Raja 'Sawai'* Jai Singh and his astronomical interests (1953). However, almost as an aside, this brilliant man had published a speculative paper in the *Indian Journal of Physics* in 1928, in which he speculated that nuclei are successively built out of protons, electrons and the nuclei of inert gases. We must remember that the neutron was still a very tentative hypothesis. In Soonawala's scheme, a lithium nucleus would be a nucleus of helium (a noble gas) plus one proton, a nucleus of magnesium would be a nucleus of neon (a noble gas) plus two protons, and so on. This was clearly influenced by Bohr's *aufbau* principle (1920), but it has in it seeds of the shell model of nuclei, later proposed by Mayer and Jensen. It was a brief flash of insight, but somehow, Soonawala never seems to have gone much beyond it.

We have described the growth of Allahabad University. Here, in 1923 came the ambitious and irascible Meghnad Saha, leaving the confines of his *alma mater*, where a financial crunch, resulting from the standoff between the obnoxious Bengal Governor and the domineering Vice Chancellor, threatened to cramp his style. Around Saha grew a group of young and enthusiastic students. However, there was another scientist of note in the newly formed Department of Mathematics. He was Amiya Charan Banerji (1891–1968), a Cambridge-trained Wrangler, fresh from his studies in Britain. Though technically a mathematician, A.C. Banerji's interests lay mostly in theoretical physics, and especially in astrophysics, where he and Saha formed a matched pair of collaborators. But he did not confine himself to the heavens. Banerji was probably the first Indian to study and develop the quantum theory of scattering and he used it in an early attempt to infer the nature of the nuclear potential from data on alpha particle scattering (1930). It was his enthusiasm that drew Meghnad Saha from his focus on statistics and thermodynamics towards nuclear physics. Their collaboration ended when Saha returned to Calcutta in 1938. Later Banerji become a well-loved Vice Chancellor of Allahabad University, where his name is still revered.

In the early 1930s, another Cambridge-returned Wrangler began to make waves in physics research. He was the mathematician Bhupati Mohan Sen, or B.M. Sen, then an obscure Lecturer at Rajshahi College (in modern Bangladesh). Sen wrote a couple of papers in *Nature* and the *Philosophical Magazine* on the intriguing subject of beta decay. At the time, just before the discovery of the neutron, the major beta decay puzzle was how electrons confined in the nucleus could acquire the energy found in beta rays. B.M. Sen tried to explain this by an application of special relativity to the energetics of beta rays. By 1933, the neutron had been found, and we find Sen, then in Calcutta at Presidency College (of which he was later to become the first Indian Principal from 1931 to 1942), publishing in *Nature* a small proof that the neutron is a Dirac fermion.

In 1927, Harold John Taylor (1906–96) came to Bombay, having earned a physics degree from Sheffield University, to teach at the Wilson College as an 'educational missionary'. He worked on a wide variety of topics, from statistical mechanics to atomic theory. Between 1933 and 1935, he was given leave to complete his Ph.D., for which he moved to the Cavendish Laboratory at Cambridge. Under Rutherford's general tutelage, he worked with James Chadwick, discoverer of the neutron. His topic was one of great current interest. As early as 1925, Marietta Blau at the Radium Institute in Vienna had pointed out that photographic emulsions smeared on to glass plates, could act as detectors of charged particles in nuclear reactions in the same way as a cloud chamber can. Her discovery met with mixed responses. Taylor's thesis at Cambridge was concerned with shooting alpha particles at nuclei and recording the events on photographic emulsions. His experiments did not earn much success, but they did get him a doctoral degree—for concluding that emulsions were not of much use. Some time before returning, however, the Rev. Taylor joined forces with the young Maurice Goldhaber. The duo were able to detect the paths of uranium fission products on emulsion, proving clearly Fermi's hypothesis of three neutrons in the final state—the reason why a chain reaction occurs. As a small aside, they found that adding some boron to the emulsion increases its sensitivity many-fold. Perhaps this success encouraged Taylor to continue working on emulsions when he returned to India. He made several studies of alpha particle scattering with his younger colleague V.D. Dabholkar. In 1954, however, Taylor left Bombay to become Principal of the Scottish Church College at Calcutta. His later life was devoted to teaching, missionary work and academic administration.

5.6 Cosmic Rays Again: Millikan and Anderson

Robert Andrews Millikan (1868–1953) was already a famed physicist when he won the 1923 Nobel Prize in Physics. His innovative oil-drop experiment and his brilliant verification of Einstein's photoelectric equation, are textbook material even today. Following these successes, in the early 1920s, this American heavyweight entered the field of cosmic ray research with all the finesse of a steamroller. With his student Ira Sprague Bowen, the resourceful Millikan was able to invent a lightweight electrometer and an ion chamber which could be sent up in unmanned balloon flights, communicating data to the ground using wireless technology developed during the Great War. This was clever, for the Europeans were still risking their lives making manned flights. Moving to Texas to make these measurements, Millikan and Bowen found much lower levels of ionisation than Hess and Kolhörster had found, and, blissfully unaware of the strong latitude effect in cosmic rays, they promptly concluded (1925) that *'the whole of the penetrating radiation is of local origin'* i.e. they come from terrestrial radioactivity.

Further measurements made with G. Harvey Cameron, in Muir Lake, 24° of latitude farther north in Canada, however, led to a 180° turn in the Nobel Laureate's opinion. In 1926, he was reporting "*... all this constitutes pretty unambiguous evidence that the high altitude rays do not originate in our atmosphere, ...and justifies the designation* 'cosmic rays'." Millikan's wholly erroneous idea that the cosmic radiation consisted of γ photons formed when protons and electrons combined to form elemental He-4 in space led him to coin the hugely evocative phrase that cosmic rays are *'the birth cries of atoms'* which form the galaxy. It immediately caught the imagination of the American Press, who lacking a hero to follow Lindbergh's 1919 aviation feat, promptly began to lionise their homegrown Laureate for his 'American discovery'—and Millikan happily basked in all their admiration. During this publicity blitz, however, nowhere did Millikan or his collaborators even mention the prior work of Linke, Gockel, Hess, Pacini or Kolhörster, leading to a furious reaction from the European scientists. In truth, Millikan had essentially reproduced all their results, and so his only original contribution to the field of cosmic rays is the name.

Millikan's idea of a cosmic gamma ray flux had always been regarded with scepticism by his fellow American Nobel Laureate, Arthur Holly Compton, but it was a Dutchman, Jacob Clay, who discovered the 'latitude effect' (1927) which proves that the cosmic rays are not photons but charged particles deflected by the Earth's magnetic field. This soon received ample confirmation from experiments carried out by Kolhörster and Walter Bothe, by Bruno Rossi and by Compton with his young student Luis Alvarez. Though Millikan refused to be convinced by this evidence, everyone else accepted it. Finally, a 1932 article in TIME magazine gently hinted that, unbelievable as it may seem, the colossus of American science might actually have been mistaken.

The first real revolution in experimental cosmic ray research came with the invention of the Geiger-Müller counter, or GM counter, in 1928. Now, instead of bulky and delicate electroscopes, there was a small and rugged instrument which could be easily sent up in balloons to count the number of γ ray photons. The wireless technology introduced by Millikan had also caught on. Henceforth, it would not be required to make risky manual flights in the cause of research—a huge relief which encouraged many to take it up seriously.

In a curious twist of Fate, the future of cosmic ray research would be determined by a student of the now-frustrated Millikan. His name was Carl D. Anderson, and his discovery of the positron in 1932 created a major sensation in the world of physics. Anderson's colleague, the Chinese student Zhao Zhongyao, had been assigned by Millikan to study cloud chamber tracks from the impact of Thallium-208 emanations on a lead plate. Zhao had actually seen positron tracks, but seems to have wrongly interpreted them as protons. Anderson performed a very similar experiment, but his source was cosmic rays. He found that the result of their collision with the lead nuclei in the plate produced tracks of a particle whose e/m ratio was equal in magnitude but opposite in sign to that of the electron. This could not be the proton, whose e/m ratio is about 1836 times smaller than the electron's, and so it must be the *antiparticle* of the electron.

The name 'positron' was actually suggested by an editor of *The Physical Review*, where Anderson published his paper. Such a particle had been suggested on theoretical grounds by Dirac in 1931, but this idea had been received with incredulity by leading scientists of the time. Now however, Heisenberg and Bohr and Pauli and Klein, all had to eat their words, for the positron was reality and their barbs against Dirac were proved to be unjustified. In fact, the undoubted existence of antiparticles roused Wolfgang Pauli out of his 'dogmatic slumber'. Within a year, he and his student Viktor Weisskopf had written down the first quantum field theory. The tentative theory of the electron field introduced in 1928 by Jordan and Wigner was now recast as the new *quantum electrodynamics,* or QED.

Abandoned by all, Millikan later recanted and accepted the consensus that cosmic rays are not photons, but charged particles, and he himself began to investigate the latitude effect, which he had actually been the first to see, but had failed to recognise. In pursuit of this, he came to India, in 1941, with his colleagues H. Victor Neher and William H. Pickering, and C.V. Raman and Homi Bhabha were very glad to receive him at Bangalore. Millikan and his team did some balloon experiments there, and their 1942 paper mentions evidence for medium-weight ions (such as silicon), as a rare component of primary cosmic rays, which they are. There was nothing more remarkable about the visit, except that it was one of the first by a prominent American scientist to India.

5.7   Particle Physics Begins in India

The first Indian paper that can be properly classified as a particle physics paper was actually written by Meghnad Saha and a young 'demonstrator' Daulat Singh Kothari (1905–93) from the University of Allahabad. As mentioned above, Saha had left Calcutta in 1923 and moved to Allahabad University, where he had slowly and steadily built up a group of young researchers around him. It was probably the enthusiasm of his younger colleague Banerji which got him interested in nuclear physics. Be that as it may, in 1927–28 Kothari and Saha addressed themselves to the famous problem of the beta particle energy distribution.

Briefly stated, the problem is this: if the beta electron arises from the two-body decay of a neutron to a proton and an electron, its energy should have a fixed value. However, observed beta particles have a wide range of energies. This was a serious puzzle in the 1920s, and even led Niels Bohr to declare that the principle of conservation of energy may be violated at the nuclear scale. Wolfgang Pauli famously solved this by postulating that 'invisible' neutrinos were created at the same time as the beta electron, and these carried away the 'missing' energy. This was later vindicated when the neutrino was discovered in 1955.

However, in 1930, when Pauli's proposal was made, not many were willing to buy his hypothesis, and, in fact, Pauli came in for quite a bit of ridicule. Saha and Kothari, however, thought up an alternative solution which was simpler and equally plausible. Noting that in most nuclides, a beta decay is accompanied by emission of a gamma photon, they claimed that the basic process was a gamma emission, where some of the photons underwent a pair creation in the electromagnetic field of the nucleus. Of this pair, one was emitted as a beta ray and the other was absorbed (together with a variable amount of energy and momentum) into the nucleus. This beautifully explains why the beta particles have varying energies, and thus *Nature* was happy to publish their letter in 1933. By then, however Kothari had left Allahabad to do his Ph.D. in Cambridge.

Saha pursued the idea for some time, but the departure of his bright young collaborator, and, more importantly, the arrival in 1934 of Enrico Fermi's theory of beta decay, which incorporated the Pauli hypothesis and explained the shape of the beta ray energy spectrum, dampened his enthusiasm. The detailed measurements of the beta energy spectrum by Franz Kurie and others, published in 1936, set the general seal of approval on the Fermi theory, and thus the Kothari-Saha theory, like so many others, fell by the wayside. However, one must acknowledge that it was a brilliant idea at the time it was conceived.

5.8   Good Intentions

The 1930s were somewhat of a roller coaster for nuclear-particle research in India. Apart from the Kothari-Saha paper, there was a brief excursus by the great Max Born, who wrote a paper with N. S. Nagendra Nath on a then-fashionable idea floated by de Broglie that the unit-spin photon is a bound state of a spin-half neutrino and its antiparticle. This is not, of course correct, but it is mirrored in the composition of a vector rho meson in terms of a quark and an antiquark. Thus do the ideas of science get recycled in different contexts.

One may wonder what Max Born, the great preceptor of Göttingen, was doing in India. It was 1933, and the anti-Semitic Nazi regime in Germany had just dismissed all Jewish professors from their positions. C.V. Raman, always on the lookout for talent, had been writing letters to several of these Jewish scientists, trying to entice them to leave faction-torn Europe and come to the peaceful environment of Bangalore, where he was heading the Indian Institute of Science (IISc). Knowing most of them personally, he fixed on two who had an abiding interest in India, viz. Erwin Schrödinger and Max Born. Schrödinger politely turned him down, saying that he could not fit himself into 'the land of the Vedas', but Max Born, whose wife Hedwiga (Hedi) was deeply interested in Hindu philosophy, did come (1935–36). Raman used his powers as Director to give Born a Readership for two years. It was then that Born worked with Raman's student Nagendra Nath and wrote that paper.

In fact, Born and his wife liked Bangalore so much that they wanted to stay there permanently. Raman, therefore, moved the Senate of the IISc to regularise Born's position by awarding him a Professorial Chair in Mathematical Physics. This should, by any counts, have been a cakewalk for someone of Born's stature, but unfortunately, Raman, in his characteristic way, had made many enemies in the Institute. It was more to embarrass him than Born (who was not too well-known outside the physics community) that the future Nobel Laureate's candidature was shot down in the Senate by a long-forgotten electrical engineer named Kenneth Aston, who forcefully argued that the Indian Institute had no need to create a chair for a 'mathematician', given that the mandate of the Institute was applied research. Twisting the knife in the wound, he stated that a premier institution like the IISc should not consider giving a place to a '*second-rate foreigner, who could not find a place in his own country*'. Max Born, who was sitting in the gallery, listening, could hardly believe his ears, and he later related "*I was so shaken that when I returned to Hedi I simply cried*".

It is India, however, who should have cried, for it was thus cheaply that an academic faction fight disposed of the great scientist and teacher who had mentored nine Nobel Laureates and would win one himself. Raman immediately appealed the decision to the IISc Council,[3] which—instead of supporting him—roundly criticised his efforts to hire a 'mathematician', essentially echoing Aston's view. Fortunately for Born, he was at that point contacted by Charles Galton Darwin (physicist grandson of the great naturalist), who was about to resign his position as the first holder of the Tait Professor of Natural Philosophy at Edinburgh in order to become Master of Christ's College at Cambridge.[4] Darwin, who understood the true worth of the German scientist, wanted Born to succeed him at Edinburgh. Born, who still retained a faint hope of staying on at Bangalore, was persuaded by Raman to accept the offer, for the harried Director could sense that his own position was under attack. Thus the hounded German professor went away to Edinburgh, and later won the Nobel Prize (1954). Sure enough, the Indian Nobel Laureate was forced by a 'Review Committee' set up

---

[3] This included, at the time, the famous *Acharya* P. C. Ray as well as Shyama Prasad Mookerjee, mathematician and son of the Vice Chancellor who had brought Raman to academics.

[4] Where the great Darwin as well as the great J.C. Bose had studied—their statues adorn the grounds of the College even today.

by the IISc Council to step down from his Directorship. With Raman's resignation ended one of the most pathetic episodes in the history of Indian science.[5]

Addressing the Indian Science Congress in 1936, Meghnad Saha put forward a theory of the difference between the proton and neutron masses based on the assumption that the neutron was a bound state of two magnetic monopoles. This was proved wrong, of course, but the idea has kept popping up ever since in different contexts—with varying degrees of success. However, this did prompt Saha to come up with his famous derivation of magnetic monopoles, which is mathematically more elegant, though intuitively less obvious, than Dirac's. We are, of course, still looking for magnetic monopoles.

From Presidency College, Calcutta, came a series of papers by Kulesh Chandra Kar (1899–1990) and his students, developing a theory that the nucleus itself has an atom-like structure, with a hard core and outer shells of alpha particles. This idea, suggested by Rutherford with his usual uncanny physical insight, may be considered a precursor of the nuclear shell model, eventually developed in 1949. The 1930s work of Kar and his collaborators, who basically used the Schrödinger equation to solve for 'nuclear wavefunctions', had elements of truth, particularly since any spherical symmetric potential leads to the same angular momentum eigenstates, but it was too naïve to describe the real nucleus. Nevertheless it may be regarded as cutting-edge research for the times.

The 1930s re-shaped the profile of science in India in many ways. The departure of Raman in 1933 to become Director of the IISc, and the joining of Meghnad Saha as Palit Professor in 1938 may be regarded as a watershed in the history of physics at Calcutta. Gone was the Nobel Laureate, with all the arrogance and aura associated with that distinction, and in his place came another brilliant man, now in his mid-forties, with very decided views on how science and technology should proceed in India. In turn, IISc gained, after a couple of undistinguished Directors, a man of great vision and immense dynamism. Despite the Born fiasco, the effect of being led by a first-class physicist soon made an impact on the physics done at the IISc. The only straight loser in this was Allahabad University, which was never again to rise to the top rank, even though it continued to produce first class students, of which Harish Chandra is a famous example.

---

[5] What makes it even sadder is that there were some other leading lights, such as the chemist Hevesey, the physicist Peierls, and even the venerable Pieter Zeeman, who were distinctly interested in Bangalore. But after learning of Born's discomfiture they shied away.

Chapter 6
Bhabha and Sarabhai

By far the most impressive work done in nuclear-particle physics by any Indian during the 1930s was done by Homi Jehangir Bhabha (1909–66), a young Parsee who was closely related to the famous industrial house of the Tatas. Before we talk of the man himself, however, it is worth describing the milieu into which he was born.

6.1 The Parsees

The Parsees form one of the innumerable communities which fled to India in the past to escape persecution in their homeland and start a new life. They are Zoroastrians from Persia, or Iran—followers of the prophet Zarathustra (Zoroaster)—and their holy book is the *Avesta*. By the year 642, Iran had fallen to the victorious Mohammedan armies, and the common people of Iran were forced to accept the stern rule of Islam—convert, or be killed. Most complied, but a small number of diehards fled the country and landed up on the coast of Gujarat, out of reach of the Caliph's cohorts. Here, over the centuries, this alien community kept their religion and traditions alive, even as the Jews of the Diaspora did wherever they went.

In 1613, the British set up a 'factory', i.e. a trading centre, at Surat, though the official sanction from the Mughal Emperor came only five years later. Soon the alien Britishers and the alien Parsees came together in close partnership. In 1661, as described in Chap. 3, the English Crown acquired the island of Bombay and by 1668, Bombay and its neighbouring islands were leased to the East India Company for an annual rent of £ 10. A trader called Dorabji Nanabhai had been the first Parsee to settle in Portuguese Bombay as early as 1640, and he was followed by a trickle of his co-religionists. The establishment of British ownership turned this trickle into a flood, and by 1675, there were 60,000 Parsees in Bombay. Most of them were engaged in

© The Author(s), under exclusive license to Springer Nature Switzerland AG 2021 87
S. Raychaudhuri, *The Roots and Development of Particle Physics in India*,
SpringerBriefs in History of Science and Technology,
https://doi.org/10.1007/978-3-030-80306-3_6

trading and commerce, but the encouragement of Governor Gerald Aungier caused many skilled craftsmen to move in large numbers to the new settlement.

The British encouraged expansion of the scope of Bombay's exports, much of which was handled by Parsee businessmen, who laid the foundations of some of the biggest fortunes to be seen in India, then and now. What made the Parsee businessmen super-rich, however, was their profitable participation in the opium trade with China. Among them was Sir Jamshetji Jeejebhoy, the first Indian to be knighted, some of whose philanthropic works we have mentioned in the previous Part. Lovji Wadia, scion of an old Parsee family of shipwrights, founded the Wadia group and built Bombay's first dock in 1736. Manekji Petit founded the first Indian cloth mills in Bombay in 1855. Ardeshir Godrej made it big by manufacturing locks in 1897. Jamshetji Tata became the great hero of Indian industry by founding his eponymous steelworks, the Taj Mahal hotel, the Indian Institute of Science and the Tata Power corporation.

It was, however, during their sojourn in the Chinese city of Canton, where the Parsee opium 'factors' had to stay in the British compound for security reasons, that this hitherto-oriental community began to acquire a taste for things of the West, such as cut-glass chandeliers, heavy furniture, porcelain and glass crockery, as well as Western clothes, music and art. They brought this liking back with them to Bombay, and soon the Parsees were the leading community in India who had adopted English ways, from their dress to their education to their use of toilet paper and knives and forks at home.

Only in their religion—to which they obstinately clung—and in their food, which was richer than European food, did any vestige of their Iranian past remain with this malleable people.

6.2   The Making of Homi Bhabha

Into this small but exclusive community was Hormusji (Homi) Bhabha born, on October 30, 1909 to Jahangir and Meherbai Bhabha. The Bhabhas were a cultured family, with Jahangir Bhabha being a successful lawyer. Jahangir's father Hormusji was a former Director of Education of the princely state of Mysore, where the Indian Institute of Science lay.[1] Homi Bhabha's mother was related to the Petit family, and his paternal aunt was married to Sir Dorabji Tata, son of the great Jamshetji. Thus, he was as near to Parsee royalty as that community could sustain, and with this background, one might imagine that he would become a millionaire, an industrialist or a successful lawyer. Fate, however, had decreed otherwise.

---

[1] As a matter of fact, Hormusji Bhabha had been a member of the Irvine Committee which induced Raman to step down as Director. However, this did not prevent Raman from offering a position to Hormusji's grandson, who bore the same name. Raman may have been cantankerous and abrasive, but science was always his first and dominating priority.

From his youth, Bhabha showed himself to be a bright student. Educated at the Cathedral and John Connon School—Bombay's best and most famous school—where he was five years junior to J.R.D. Tata, the future head of the entire Tata group, he cleared the Senior Cambridge Examination at the age of 15 and was enrolled at the venerable Elphinstone College for a bachelor's degree in science—for which he already showed a particular aptitude. The patriarchs of the family—his father Jehangir and his uncle Sir Dorabji—had by then put their heads together and decided that this bright young lad was just what the Tata Iron and Steel Works needed as an engineer and a metallurgist, who could be groomed to take over the management in the years to come. For higher studies, then, the family was happy to sponsor the young man to go to Cambridge University—the Mecca of science in those days—where he would take the Mechanical Sciences Tripos, a course in which he would learn a combination of what we now call mechanical, civil and metallurgical engineering. On his return, a great future awaited him in the Tata empire.

With all the optimism of youth, the handsome young sprig sailed off to Europe in 1927, from where he wrote dozens of long letters to his parents, his beloved younger brother Jamshed and his maternal aunt—his favourite 'Coomy auntie'. From these scarce-decipherable letters one can gather that his studies sat lightly on the brilliant young Parsee lad, and that he found plenty of time to indulge his other interests—painting, music and the theatre. He made friends easily—irrespective of gender—and many of these friendships lasted as long as he lived. Thus, Homi Bhabha loved every moment of his sojourn in the ancient Gonville and Caius College of the University of Cambridge. He played music, created sets for plays, painted pictures and enjoyed the intellectual—and not so intellectual—discussions over table. He took to the milieu as a duck to the water. This was largely due to his westernised upbringing. Not for him were the agonised efforts of a Ramanujan or a Gandhi to adjust to the *sahib*'s way of living.

When the young Indian arrived at Cambridge, theoretical physics was just beginning to go through the biggest revolution it had seen since the Renaissance. In Germany, Werner Heisenberg had enunciated the Uncertainty Principle (1925), and in Switzerland, the equally brilliant Ernest Schrödinger had created wave mechanics (1927). A tall, reticent young professor by the name of Paul Dirac was to go further and formulate a relativistic equation which explained for the first time why the electron carried a mysterious quantum of angular momentum called spin (1928). At the same time, the famed Cavendish Laboratory was being headed by the leonine Lord Rutherford, and his disciples Pyotr Kapitza, James Chadwick and John Cockcroft were making great advances in nuclear physics. It was impossible to be at Cambridge in those exciting days and not become subject to the overwhelming attractions of the quantum revolution.

Soon, the young Bhabha was writing to his father, saying how he was desperate to do theoretical physics and not engineering. This is a scenario oft repeated in present times, but alas! very few Indian parents react as the hitherto-stern Jehangir Bhabha did. He gave his permission to his recalcitrant son to enrol for the Mathematical Tripos at Cambridge, *provided* he finished the Mechanical Tripos first. Thus, suitably chastened, Homi Bhabha put his head down and cleared the Mechanical

Tripos in 1929. He then enrolled for the notoriously-difficult Mathematical Tripos and cleared that with flying colours in 1932, winning the prestigious Rouse Ball Travelling Studentship in Mathematics. He was something of a protégé of Dirac, yet he enrolled with Professor Ralph H. Fowler for his Ph.D. Fowler was Cambridge's greatest midwife of theoretical talent—fifteen of his students became Fellows of the Royal Society and three of them went on to win Nobel prizes.[2] Under his gentle supervision, the young Parsee plunged into the study of cosmic rays, the most intriguing mystery of the early decades of the twentieth century.

But what are these mysterious emanations from the cosmos, which cause the rarefied air in the upper atmosphere to ionise and become a very tenuous plasma? After a hundred years of research, we now have a pretty good idea of their origin and nature. Cosmic rays consist of very high energy particles emitted when distant events of unimaginable violence take place in the Universe, such as supernovae (the blowing up of giant stars), quasars (collision of entire galaxies) and active galactic nuclei (super-massive black holes swallowing up entire galaxies). A very high proportion of them come from our own Sun, where storms (called sunspots), enormous explosions (called flares) and spewing out of hot gas (called protuberances) are happening all the time. Most of these particles are simply protons or alpha particles—nuclei of hydrogen and helium stripped of their electrons—but some are more exotic particles, such as electrons, neutrinos and the nuclei of heavier atoms. What we detect on the ground or in balloons, however, is not these primary cosmic rays, but secondary particles emitted when these hit the nuclei of nitrogen and oxygen atoms which make up the air. The latter include neutrons, pions, muons, electrons and positrons, as a well as a few more exotic ones like kaons and heavy baryons. However, all this was not known when Bhabha began his researches. It was just known that some form of ionising stuff was hitting the Earth's atmosphere and that it came from outside the Earth.

Within one year, Bhabha had published his first paper, a masterly discussion of the absorption of cosmic rays as they pass down through the atmosphere. This appeared in the *Zeitschrift für Physik*, a prestigious physics journal published from Germany. In the same year, however, Carl Anderson discovered the positron—the antiparticle of the electron—in cosmic rays, and this discovery intrigued Bhabha among many others. His next two years were spent in theoretical studies of electron and positron production and absorption in cosmic rays, leading to his doctoral thesis which he submitted in 1935. On the strength of this, he won an Isaac Newton Fellowship for the next three years—today it would be called a postdoctoral fellowship—which enabled him to stay on at Cambridge and also go to the European continent, where he was able to work with stalwarts like Niels Bohr, Wolfgang Pauli and James Franck. His 1933 paper in the *Zeitschrift für Physik*, entitled "*On the absorption of cosmic rays*" acknowledges discussions with the great Wolfgang Pauli in the Introduction itself.

It was now that Bhabha produced one of his two most famous pieces of scientific work.

---

[2] Fowler's other claim to fame was as the son-in-law of the great Ernest Rutherford, who would often refer the theoretical side of his researches to him. This led to a very fruitful partnership.

6.3   Bhabha and the Antiparticle Hypothesis

Inspired by his hero Dirac, Bhabha had started thinking deeply about the electron and the new QED ideas. The starting point was Dirac's theory of the electron, enshrined in the Dirac Equation, which he wrote down in 1928. This was a successful attempt at finding a quantum theory of the electron which would be compatible with Einstein's special theory of relativity, for by that time everyone (except Nazis and a few mavericks like K. C. Kar) was convinced of the correctness of relativity. Unfortunately, in relativity, there is a possibility for the kinetic energy of a free particle to be negative. If this really happens, we get what is called a catastrophe, for energy tends to be degraded, and there is nothing to prevent the particle from going down towards a negatively infinite energy, emitting an infinite amount of energy in the process. This is not an issue in classical physics, for once the energy is positive, it will remain positive, and we can just say that there are no particles which have negative energies. In quantum mechanics, however, it is possible for a particle to make a discrete *jump* in energy from the positive to the negative, triggering the catastrophe.

Dirac found a makeshift solution to this problem for the electron. The Dirac equation works only for particles with spin one-half (in units of the reduced Planck's constant $\hbar$), such as the electron (and also the proton and neutron). Such particles are known to obey Pauli's Exclusion Principle, that no two particles can go into the same state, defined by the same set of quantum numbers. Dirac, accordingly, assumed that all the negative energy states already contain electrons, which are—for some unknown reason—invisible. Thus, it would not be possible for a positive-energy electron to jump into a negative energy state. The catastrophe was averted, but the opposite could happen, viz. a negative energy electron could jump into a positive energy state, suddenly appearing as a new, visible electron. Since electric charge is conserved, the 'hole' left in the sea of negative electrons would appear like a positively charged particle. Dirac initially suggested that it might be the proton, but it was soon realised that the conservation of energy and momentum would ensure that the mass of this 'hole' must be the same as that of the electron. By 1931, Dirac was bold enough to say that it would be a new kind of particle—a kind of positive electron.

This bizarre theory earned Dirac more brickbats than plaudits. Some of the most caustic comments came from Heisenberg and Pauli, the Young Turks who were at the forefront of building the new ideas of quantum mechanics. Heisenberg, in particular, described it as "... *learned trash which no one can take seriously*" and declared that "*The saddest chapter of modern physics is and remains the Dirac theory.*" Pauli, in a more jocose vein, started signing his letters as "*Yours (drowned in Dirac's formulae), W. Pauli*". Even the normally sedate Niels Bohr is on record for having said "*Oh, but that is no theory. That is, not a theory that one can believe in.*".

When the positron was actually discovered by Anderson, as described in the previous section, all these naysayers of Dirac's theory were silenced. Being first-class physicists, however, they quickly recovered from the shock and addressed themselves to developing a better understanding of the new phenomenon. For relativity applies as much to bosonic particles i.e. those with zero or integer spin, and there the Pauli exclusion principle does not hold. Thus, even if one postulated a boson sea, like Dirac's electron sea, there would be nothing to prevent a positive-energy boson hopping down to a negative energy state already occupied by an identical boson, and therefore, the catastrophe could not be prevented.

It was Pauli who solved this problem, and his solution was nothing short of brilliant. In his epoch-making 1933 paper with Weisskopf, it was assumed that there are two types of particles with identical mass and other quantum numbers equal and opposite, all having positive energies. The opposite sign in the quantum numbers balances the negative sign in the kinetic energy of one class, rendering their overall energy positive. Thus, there is no negative energy sea, and no holes. There are only particles, and these *anti*-particles, which have opposite quantum number. Since relativity allows the inter-conversion of matter and energy, it is always possible for a particle and an antiparticle to coalesce and *annihilate*, creating pure energy. It is also possible for a quantum of pure energy, like a photon, to get converted into a particle-antiparticle pair. Thus, a high energy particle is never a single particle, but always has some of its energy popping in and out of the vacuum as particle-antiparticle pairs. This model of an elementary particle is called a *quantum field theory*. Soon, it appeared that it could be applied to an electron–positron theory as well, and also to photons, i.e. the electromagnetic field. Combining the last two gave *quantum electrodynamics* or QED, the quantum analogue of Maxwell's classical electrodynamics.

However, the euphoria over the invention of quantum field theory and QED was short lived. It had, been applied to electron–electron scattering quite successfully by Neville Mott (non-relativistic case) and by Christian Møller (relativistic case). However, when higher orders in perturbation theory (the staple calculation tool of quantum mechanics) were computed, the results came out to be infinite, or, in technical language, singular. Top scientists like Heisenberg, Bohr and Kramers, started puzzling over this, and trying to find an understanding, but most others thought it a good reason to flatly reject the whole theory of antiparticles. The positron, according to them, was just a new particle which has the same mass as the electron and a positive charge. Having the same mass was not such a big thing, for the proton and neutron have the same mass (the difference is only 0.12%). If every particle has an antiparticle, reasoned these hard-headed empiricists, where were they? Why did only the electron have an antiparticle? Surely the whole concept was flawed.

This was the fundamental issue which Homi Bhabha addressed himself to. It is characteristic of the man that his approach was not mathematical (though he knew as much mathematics as anyone else) but phenomenological. And so, in 1936, Bhabha published in the *Philosophical Magazine*, a paper entitled *"The scattering of positrons by electrons with exchange on Dirac's theory of the positron"*, in which he took up the question of a positron scattering from an electron (or vice versa) in the process which is now universally called *Bhabha scattering*.

In a QED process, a positron can radiate a virtual photon carrying away energy and momentum, and recoil in the process. The photon can then be absorbed by an electron, which recoils in turn. The recoiling electron and positron pair will appear to have scattered from each other with transfer of energy and momentum from one to the other, exactly as in a classical scattering process. Such a process would also occur between the electron and any other charged particle, such as a proton or an ion or a heavier nucleus and so it would not depend on the positron being the antiparticle of the electron. In the modern language of Feynman diagrams (which had not yet been invented), this is a 'scattering diagram' or a 't-channel process'..

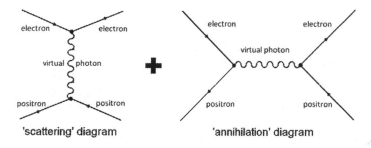

There is, however, another QED configuration which is possible only if the positron is the antiparticle of the electron—they can collide and annihilate to form a virtual photon which then decays into a new electron–positron pair. Obviously this will not happen with a proton or an ion. Again, in the language of Feynman diagrams, this is an 'annihilation diagram', or an 's-channel process'. What Bhabha realised is that these two processes will undergo a quantum interference, as a result of which the forward-scattering would get favoured over the backward scattering, whereas in the absence of this interference both would have been the same. This is analogous to the difference in intensity between rainbow and glory scattering, though for rather different reasons. Since such a difference had already been observed, Bhabha concluded that the interference did happen and therefore the annihilation process was occurring; *ergo* the positron must be an antiparticle!

This brilliant paper instantly brought the scales down heavily in favour of the existence of antiparticles, and by implication, of the correctness of quantum field theory. It also gained Bhabha instant fame, and henceforth he was treated on terms of equality with the who's who of contemporary physics. Upon the expiry of his Isaac Newton Fellowship in 1937, he was awarded one of the curious fellowships that exist only in England, called the Senior Studentship of the Exhibition of 1851,[3] which enabled him to live and work in Cambridge some more time.

The understanding of the singularities in QED, when it came, turned out to be the theory of renormalisation, still one of the more abstruse concepts in physics, which very cleverly uses the machinery of quantum field theory to push singularities into non-measureable quantities in the theory, leaving all the experimentally measureable quantities finite. It leads, however, to a variation in physical constants such as the

---

[3] The Great Exhibition of 1851 was an exhibition of industrial and scientific inventions held at London's Hyde Park and may be considered the world's first trade fair. Among other things, it made a huge profit from selling tickets, creating a corpus which was—and still is—used to fund scientific and industrial research.

electron charge with increasing energy—something which has subsequently been accurately verified in hundreds of experiments.

6.4   Cosmic Ray Showers

The Nazis' coming to power had been accompanied, as we have seen, by an exodus of Jewish scientists from Germany. Among them was a youngish, but already famous, theoretician, called Walter Heitler. Heitler was a skilled exponent of second quantisation, and was soon (1936) to compose his famous "*Quantum Theory of Radiation*", which became a classic textbook in the subject for decades. Now Heitler was a refugee in England, with a temporary position at the Wills Laboratory of the University of Bristol, but he would frequently come to Cambridge. In 1934, he published a very important work with Hans Bethe—discoverer of nuclear fusion—who was still in Germany, but was soon to escape to America. This had to do with the subject of cosmic rays, which was still Homi Bhabha's pet subject.

It was well known that high energy electrons form part of the cosmic rays and are regularly detected at ground level. This means that they pass through 8–10 km of air of increasing density as they move down from the top to the bottom of the atmosphere. Bethe and Heitler set out to calculate—using the techniques of the newly-developed quantum electrodynamics—how far an electron can travel through the air, full as it is of the heavy nuclei of nitrogen and oxygen. Their estimate was about 2 km, which is far too small to explain the observation. If this were to be accepted, hardly any cosmic ray electrons would reach the ground, and not the large numbers that actually do. It was this puzzle that Bhabha discussed with Heitler. Unlike many of the senior scientists of the time, who simply dismissed the Bethe–Heitler calculation as wrong, because they did not believe in quantum field theory, Bhabha took it very seriously.

The clue came from an observation published by Patrick Blackett and Giuseppe Occhialini from London in 1931. Like many others, they studied photographs of the tracks made by cosmic ray particles in a cloud chamber. But among the usual formations, they found a few cases where the energy of the initial particle entering the cloud chamber and hitting some nucleus was enough to cause the emission of another fast particle, which in turn hit another nucleus and caused emission of more particles, and so on, till all the energy got dissipated. Thus, instead of a single particle, there was a shower of particles, or in formal language, a cascading process. In fact, this is exactly how a landslide or an avalanche begins—a rock or a lump of snow at high altitude falls, dislodging more rocks or mud or snow, and this in turn rolls down, dislodging others, until a whole hillside plunges downwards in an overpowering rush. For cosmic rays, however, the Blackett-Occhialini finding was dismissed by most contemporaries as an interesting, but rare phenomenon.

Not so by Bhabha and Heitler. They were able to combine the Bethe–Heitler result with the Blackett-Occhialini finding in a very clever way. High energy electrons in the upper atmosphere, they claimed, were indeed absorbed within 2 km, as the Bethe–Heitler calculation demanded. But in losing energy they would knock out electrons, and other particles, from the nuclei of nitrogen and oxygen atoms in the air, and these would create a cascading effect, exactly as was seen in the Blackett-Occhialini cloud chamber. In fact, not just electrons, but also protons and other high energy cosmic rays, would all start such cascades from the upper atmosphere, resulting in what is now called a *cosmic ray shower*, or, among cosmic ray physicists, an 'air shower'. The electrons detected at ground level are not, then, coming from the top of the atmosphere, but are created a kilometre or so high, well within the limit set by Bethe and Heitler.

Bhabha and Heitler published their idea in the *Proceedings of the Royal Society* in 1936. In those days, ideas percolated much more slowly than in these days of internet, and so J. Robert Oppenheimer and his student Carlson, across the Atlantic, independently came up with much the same idea, which they published in the *Physical Review*, the journal of the American Physical Society, in 1937. But Bhabha and Heitler had beaten them to the punch. The theory of cosmic ray showers may be found today in any textbook, and it is always referred to as the Bhabha-Heitler theory.

6.5   Heroic Days

By 1937–38, Homi Bhabha was already a scientist of note, and he would be invited to the same conferences with contemporary greats like Dirac, Chadwick, Heisenberg, Pauli and Peierls, not to speak of living legends like Rutherford, Bohr, Mme. Curie and Max Born. During this short period, he made at least three observations which were far ahead of his time.

In 1935, Hideki Yukawa, the great Japanese theorist, had suggested that the strong nuclear force between protons and neutrons is mediated by a particle of intermediate mass and zero spin. Yukawa called this the 'mesotron'. This particle does indeed exist, and is today called the pion. It can come in both charged and neutral varieties. Most cosmic ray physicists of the time believed that the so-called 'hard' component of the cosmic rays, which penetrated to the ground, consisted of Yukawa's mesotrons. It was Bhabha who first pointed out in 1936 that if mesotrons were to have the properties predicted by Yukawa, then they could not travel more than a few centimetres in air, and certainly not through the entire atmosphere to reach the ground. The 'hard' radiations must, he reasoned, be a different particle, not known before, which had a similar mass, but behaved more like an electron than a Yukawa particle. Indeed, in 1937, Carl Anderson and his student Seth B. Neddermeyer discovered a new particle in cosmic rays which does behave like an electron, except that it is about 206 times heavier, and decays to an electron in about 2 microseconds. This particle was eventually named the μ meson, or muon.

The second of Bhabha's contributions was something which was revolutionary in its time, but has become such a commonplace today that it is a standard exercise for every physics student. Muons are produced in the upper atmosphere like electrons, by the decay of pions formed in the impact of cosmic rays on the nuclei of nitrogen and oxygen in the air. Since muons decay in about 2 microseconds, then, even if they could travel at the speed of light, they can traverse at best a distance of around 600 m. How then do they reach the ground level, 8–10 km down, where they are regularly detected? The Bhabha-Heitler theory will not work here, as cascades rarely have enough energy to create muon pairs. It was here that Bhabha showed great perspicacity in pointing out that the relativistic time dilation experienced by a muon moving at velocities close to the speed of light will increase its lifetime more than 20 times, which is enough to allow it to reach the ground before decaying. This has become a standard textbook exercise today, but few remember that it was first pointed out by the young Indian. This feat is all the more significant because in the 1930s, there was a vociferous group of German scientists—enemies of Einstein simply because he was Jewish—proclaiming that relativity was all wrong. Here was a direct experimental proof of relativity to throw in their face.

The pion predicted by Yukawa was not discovered till 1947 as will be related in the next chapter. However, Homi Bhabha was quick to point out (1937) that if Yukawa's theory were correct, there should also be a spin-1 counterpart of the pion. The modern name for this is the rho (ρ) meson, and it too, as Bhabha said, comes in both charged and neutral varieties. However, it was not till 1961—almost a quarter of a century later—that the rho mesons were actually discovered at the Brookhaven National Laboratory (which was set up only after the War) in America.

Towards the end of the 1930s, as Europe and the world lurched towards the carnage that we call the Second World War, Homi Bhabha was making a name for himself among the up-and-coming young scientists of his time. In a host of now-yellowed letters, photographs and conference proceedings, we get a glimpse of the young savant walking and talking as an equal with giants like Einstein, Bohr, Pauli, Dirac and Yukawa. Perhaps his Westernised Parsee upbringing helped, but no Indian before him, neither Ramanujan, nor S.N. Bose, nor Meghnad Saha, not even Nobel Laureate C.V. Raman, ever achieved the degree of respect and acceptability among European scientists that Homi Bhabha commanded in those halcyon days. For all purposes, he was a Cambridge scientist, to be counted with the best and most brilliant. Alas! this was not to last.

6.6   Reality Check

On September 3, 1939, the Second World War began. It was to rage for the next 4,024 days in Europe and end only with the atomic bombing of Japan. As the clouds of war gathered over Europe, Homi Bhabha came under tremendous pressure from his family to come back to India. In any case, Bhabha's scholarship of the 1851 Exhibition was also due to come to an end in August 1939. So it was with some misgivings that he sailed back to India in that summer. The storm broke over Europe while he was still on vacation with his family.

There was no question of going back to England. It would not be till 1946 that Bhabha would see the ivy-covered walls of his beloved college again. In any case, Bhabha was now 30 years old, and it was time to seek a more permanent position.

Forced, therefore, to seek a job in India, Bhabha wrote to a few friends and acquaintances, seeking a position. With his Cambridge record, he was a great catch. A deluge of offers followed, but only three of them seemed to have cut any ice with him. One was from Allahabad University where the Physics Department offered Bhabha a professorial chair. This was a generous offer for a 30-year old, but no one there was interested in cosmic rays. Moreover, there would be a fairly heavy teaching load. Bhabha hesitated.

More congenial, from an academic point of view, was an offer from the Indian Association for the Cultivation of Sciences in Calcutta, where Meghnad Saha was now the President. Calcutta was the hottest centre of Indian physics then, but the only person interested in cosmic rays was the very senior D.M. Bose, a true-blue experimentalist with little interest in theory. As a rich Parsee and a declared Anglophile, Bhabha would never be on the same cultural plane with his Calcutta peers, who were all rather *swadeshi*-minded. No, he would not go to Calcutta after all.

The offer which finally won over Bhabha came from the Indian Institute of Science. Raman, no longer Director, but undaunted in his search for talent, offered Bhabha a Reader's position at the Institute, with a free hand to build a research group in cosmic ray physics. But where was the money to build such a group to come from? All Raman could get Bhabha was his salary and two rooms in the beautiful building which housed the Institute. But Bhabha himself had the key to the golden staircase. He could directly tune in to his late uncle's commercial empire, whose millions were funding the Institute anyway. And thus it was with a generous grant from the Sir Dorab Tata Trust, which came in 1942, that Bhabha was finally able to set up his dreamed-of cosmic ray group at Bangalore.

Since his childhood, Bhabha had only seen the brighter side of life. It was only at Bangalore that the pampered Parsee genius came face-to-face with the realities of doing top-class research in a colonised and impoverished country. For the first time, he would have to tailor his research to limited funding and to interact with mediocre scientists, meddlesome bureaucrats and an overweening Director—things which were (are?) mother's milk to more home-grown scientists. Many a bright 'foreign-returned' spark has been extinguished in such circumstances. But not so Homi Jehangir Bhabha. He was made of sterner stuff. If the right environment was missing, he would create it. And that is what he strove to do for the next quarter of a century.

6.7   Balloons and Brainteasers

The Indian Institute of Science had been a sleepy place till Raman come in 1933 and shook it out of its peaceful repose. When the din and clangour of Raman's Directorship was over, the Faculty settled down again, under the easygoing J. C. Ghosh, to 'the noiseless tenor of their way." But Raman had not gone, and his new protégé was to set up a furious pace of work during the next half-a-dozen years which he spent there.

Few scientists in the modern era have excelled in both theory *and* experiment. Enrico Fermi was the most famous of them. In recent times we have Kip Thorne, a trained theorist who won the Nobel Prize for a great experiment. Homi Bhabha belonged to this rare class. His work at Cambridge had been purely theoretical, though he had kept abreast of the developments in experimental physics—especially as regards cosmic rays. There were plenty of excellent experimental physicists around, and he only had to keep his eyes and ears open to get all the news. Now at Bangalore, in wartime, there was complete silence. Most of the experimentalists he knew were involved in the war effort and, in any case, letters took ages to reach. Nothing daunted, Bhabha decided that he would be his own experimentalist. He would fly his own balloons, and get his own data. It was here that the Engineering Tripos he had done so reluctantly came in handy.

Setting up the Cosmic Ray Unit at the I.I.Sc., Bhabha found an able assistant in Ranjan Roy Daniel (1923–2005), a Raman protégé whom the great man diverted to help in the cosmic ray project. An experienced collaborator was Prof. S. V. Chandrasekhar Aiya, of the Electrical Engineering Department. Two Bose Institute-trained experts who came to join the group were R.L. Sengupta and M.S. Sinha. The latter had learned to build a cloud chamber from D.M. Bose (Sengupta had also worked with Blackett in London) and they carried out ground-based work with the one they constructed at Bangalore. But Bhabha's creativity (literally) soared to great heights in the balloon programme.

Bhabha's first concern was to separate out the 'soft' and 'hard' components of the cosmic rays, which he did by inventing what came to be known as 'Bhabha counters'. These consisted of successive layers of absorbing materials, interspersed with counters in coincidence. So successful was this device that Bhabha and his collaborators could tell primary cosmic ray showers from secondary ones. Initially, they manage to get them carried in aeroplanes to great heights, of 15,000 ft, then 30,000 ft and finally 40,000 ft (a modern jet airliner flies at 35,000 ft). But friendly aeroplanes grew scarce as the war reached Indian shores. At this point, Bhabha and his team developed flight balloons. These could stay afloat much longer and collect much more data. And so they did.

Comparison of their result with those of Schein, Jesse and Woller at Chicago showed that the hard component of cosmic rays showed almost no latitude effect, but the soft component did. This was consistent with Bhabha's own hypothesis that the hard component was mostly muons created in the atmosphere, whereas the soft component was primarily protons from the Sun, as Blackett had guessed. Obviously, in their much longer flight, the protons would have time to be deflected much more by the Earth's magnetic field than the muons, which were essentially produced locally in the upper atmosphere.

At the same time as his balloon studies, Bhabha gathered around him a group of young theorists and phenomenologists. A 1940 paper in the *Proceedings of the Indian Academy of Sciences* discusses the possibility of Dirac fermions of charge greater than unity, something which seemed, in those early days of QED, to have problems. Bhabha pursued this for a couple of years and then moved to integral spin. In B.S. Madhava Rao (1900–87), who had just joined the Central College,

Bangalore with a D.Sc. in Applied Mathematics from the University of Calcutta, the mathematician in Bhabha found a kindred soul to bounce ideas off. With the arrival of D. Basu, a former student of M.N. Saha, Bhabha had a small group to discuss Group Theory—then something of a novelty in quantum theory. But his main concern at the time was still the Dirac equation. The relativistic electron theory of Dirac assumes a fixed electric field background and does not take into account the effect of radiation reaction. Dirac (1938) was able to develop a theory which includes radiation reaction, but in this work the electron spin was taken to be zero. Bhabha and H. C. Corben were able to generalise this theory to spin-½ particles, inventing the 'Bhabha-Corben equations'.

Corben was at Cambridge and so Bhabha completed their collaboration by the slow and painful method of exchanging letters. Likewise he continued to collaborate with Heitler. They found that the vector mesons (rho mesons) postulated by Bhabha in 1937 could scatter from nuclei only if there existed 'nucleon isobars' of higher spin. These were, in fact, found by Fermi in 1952. But it was not with Heitler, but with his own student S.K. Chakravarty, that Bhabha continued his work on cosmic ray showers. The initial Bhabha-Heitler paper had considered the development of air showers, but had not taken into account the energy losses due to collisions of the shower particles with the air. This was remedied in the work with Chakravarty, where a series was developed to calculate the cross-section and summed by an ingenious technique.

With hindsight, perhaps Bhabha's greatest contribution to theoretical science from his Bangalore days was to train the young Harish Chandra, and send him to Dirac at Cambridge, from which ultimately emerged Harish Chandra's classic works on group representation theory.

It is amazing that one man could have started from two empty rooms and in six years' time built up a group with so much activity, and produced so many first class results. One must take one's hat off to Bhabha's energy and creativity. And yet, none of this Bangalore work had quite the originality and significance of the Cambridge years that preceded them. Bhabha and his team did pioneer balloon flights with cosmic ray detectors in India, and made many good and useful measurements of cosmic rays using the advantages offered by India's low latitudes, but they were not fortunate enough to make new discoveries. The poorer quality of equipment available in India at the time could well have been a reason. The absence of discussions with top experts like Rutherford, Wilson, Cockcroft, Carmichael and others may have been a more potent reason.

Perhaps the burden of doing both theory and experiment was a bit too heavy, even for Bhabha's broad shoulders, though he revelled in it. One might wonder what Bhabha might have done had he gone back to Cambridge after the War and devoted himself to the theoretical physics which was really his forté. However, it must be noted that he would have been 36 years old and, with Indian independence in the offing, it would have not have been easy to get a position on the faculty there. In any case, all this is idle speculation, for Fate had marked out a different role for Homi Jehangir Bhabha.

For his cosmic ray work, Homi Bhabha was nominated for the Nobel Prize five times. Interestingly, every time it was by the French mathematician Jacques Hadamard and not by any of his Cambridge cronies.[4] However, competition was tough in those days, and he lost out, in successive years, to John Cockcroft and Ernest Walton (particle accelerator), Frits Zernike (phase contract microscope), Max Born (quantum mechanics) and Walter Bothe (coincidence technique), Willis Lamb and Polycarp Kusch (tests of QED) and finally, the transistor trio, John Bardeen, Walter Brattain and William Shockley. Had he been blessed with the extreme longevity of some of the current prizewinners, he may have eventually got the Prize. But even there Fate intervened.

6.8   The Creation of TIFR

In 1944, the Inspector General of Education of Bombay offered Bhabha a professorial chair at the Royal Institute of Science, his own *alma mater*. Bhabha's father had died shortly before that and there were strong family reasons for returning to Bombay. However, Bhabha politely declined. He already had a professorship at Bangalore, and it would be difficult to move his whole group to the Royal Institute. Moreover, there would be a teaching load.

In fact, Bhabha had already gone some way down a different path. In 1943, he had written to his cousin, J.R.D. Tata, then the youthful head of the Tata business empire, that he wanted to stay in India, but found the lack of a proper atmosphere and adequate funding quite daunting. The reply from Tata was encouraging. "*...If you and/or some of your colleagues in the scientific world will put up concrete proposals backed by a sound case, I think there is a very good chance that the Sir Dorab Tata Trust...will respond. After all, the advancement of science is one of the fundamental objectives with which most of the Tata Trusts were founded, and they have already rendered useful service in the field. If they are then shown that they can give still more valuable help in a new way, I am quite sure that they will give it their most serious consideration.*"

There is a story behind this. When Jamshedji Tata was trying to found his famous steel plant, he was directed to a region in the Central Provinces (now in the state of Chattisgarh), where there seemed to be high grade iron ore of the haematite variety. Unable to find a good laboratory in India to test these, Tata had taken ship to England in 1893, accompanied by some sacks full of rocks from the site. On the ship, he struck up a friendship with *Swami* Vivekananda, then travelling on the ship to what was to be his great triumph at the Chicago Parliament of Religions. The *Swami* suggested to him that instead of rushing halfway across the world for this purpose, he should set up laboratories in India which could make such analyses. The words struck a deep chord in the Parsee entrepreneur, and thus he made an Institute of Science one of the four major planks of his *swadeshi* programme. However, Lord Curzon immediately

---

[4] They did, however, make him a Fellow of the Royal Society in 1941. Dirac was the proposer.

shot down the proposal (1898), and the pioneer died without seeing it fulfilled. It was left to his sons, Sir Dorab Tata and Sir Ratan Tata to actually create the Indian Institute of Science at Bangalore, where Bhabha was working in 1943. Thus, the Tata group's commitment to science ran deep and J.R.D's letter was in the same spirit.

This was to be put to test very soon. Bhabha wrote up a masterly proposal (March 1944), from which it is worth quoting verbatim. *"There is at the moment in India no big school of research in the fundamental problems of physics, both theoretical and experimental. There are, however, scattered all over India competent workers who are not doing as good work as they would do if brought together in one place under proper direction. It is absolutely in the interest of India to have a rigorous school of research in fundamental physics, for such a school forms the spearhead of research not only in less advanced branches of physics, but also in problems of immediate practical application in industry. If much of the applied research done in India today is disappointing or of very inferior quality it is entirely due to the absence of a sufficient number of outstanding pure research workers who would set the standard of good research and act on the directing boards in an advisory capacity."* The Tata group responded rather fast. The trustees met to consider the proposal in April 1944. The new 'Tata Institute of Fundamental Research' started functioning from June 1, 1945 with a generous endowment of 80,000 rupees for the first year. Bhabha would never go back to Cambridge, except as a visitor. Instead, he was Director of his own Institute, which he could build up as he liked, and that is just what he did.

Bhabha's Institute was to be located—no surprise—in his home city of Bombay. In the beginning it was located in the sprawling bungalow 'Kenilworth', owned by his aunt Miss 'Coomey' Panday. This frail old lady lived in a small part of the bungalow and let her favourite nephew use the rest to fulfil his dream. After some time, the Institute grew too large for the bungalow, and moved out to the sprawling Tudor-style manor house that was the Old Yacht Club of Bombay. Even that proved inadequate, however, and eventually the Institute moved to its present location in the Bombay cantonment area, with its own specially-designed lavishly-appointed building (1962).

It is worth relating that the very first scientist sounded out by Homi Bhabha for his new Institute was Subrahmanyam Chandrasekhar, his old Cambridge friend, who was then at Chicago. To him, Bhabha wrote *"It is our intention to bring together as many outstanding scientists as possible in physics and allied lines so as to build up in time an intellectual atmosphere approaching what we knew in places like Cambridge and Paris..."* Despite this clarion call, however, Chandrasekhar vacillated, exchanged many letters, and eventually elected to remain in Chicago. Bhabha was more successful in luring away D.D. Kosambi, a Harvard-trained mathematician, from the Fergusson College at Pune, and R.P. Thatte, an electronics expert from the S.P. College, Pune. The Institute started with Bhabha and Kosambi as Professors. Thatte was the very first Ph.D. from TIFR, which he completed under Bhabha's supervision. Soon a whole galaxy of talent gathered around Bhabha in the new Institute. The overall story of TIFR, however, does not concern us here, except the cosmic ray and particle physics research, which will be described presently. It is important to note, however, that Bhabha always gave great importance to maintaining an ebb and flow of international visitors to his Institute. Some of the big names which visited in the first two decades include physicists P.M.S. Blackett, P.A M. Dirac, Hannes Alfvén, Murray Gell-Mann, Felix Bloch, John Cockcroft, J.D. Bernal and C.F. Powell and mathematicians Carl Siegel, Andre Weil, Laurent Schwartz, Norbert Wiener, Harish Chandra and Paul Erdős.

As for the brilliant founder himself, just 11 days after India won her independence from the British *raj*, he was made Chairman of a 'Board of Research in Atomic Energy'. This was in everyone's minds after the hammer blows which Hiroshima and Nagasaki had wrought on the world's conscience. Fermi's nuclear reactor and the promise of cheap and unending energy looked like a dream of an industrial utopia. It was natural to put India's most dynamic scientist in charge of this. Out of this came the Atomic Energy Commission (1948), which set India along the road to the nuclear club. From now on Homi Bhabha was increasingly an institution builder and a science manager, cutting a wide swathe through the scientific community until his untimely death at the age of 57, in an air crash. Some elegant theoretical work was still to come from this brilliant mind, but somehow, it never managed to reach the heights of his phenomenological work at Cambridge. However, if TIFR—and indeed the whole of post-independence Indian science—has fallen short of Bhabha's dream of a new Cambridge and a new Paris, surely it cannot be blamed on that extraordinary man.

6.9   The Young Sarabhai

Homi Bhabha was not the only scientific luminary in the sky of newly-independent India. Vikram Ambalal Sarabhai (1919–71), ten years his junior, was a man of equal talents and an even greater institution builder. He left behind him a legacy and a reputation that equalled the Parsee savant in many ways, and surpassed it in some, though, to be fair, none of his scientific work was as fundamental as Bhabha scattering or the Bhabha-Heitler theory.

We have related how the Parsees migrated to Gujarat after they fled from Islamists in conquered Iran. Gujarat itself came under Islamic rule when it was conquered by the fierce Alauddin Khalji in 1297 and incorporated into the Sultanate of Delhi. Regaining its independence after a century of so, Gujarat remained under its own Muslim rulers (mostly of Turkish origin) till it fell under the sway of the Mughals. When, in turn, the Mughal empire weakened, Gujarat was wrested away by the Marathas, one of whose powerful captains, Damaji Gaekwad, made it his own fiefdom. Eventually, the Gaekwad, who ruled from his capital at Baroda (Vadodara), accepted the suzerainty of the British, and matters remained thus till the Gaekwad and other minor rulers all acceded to the free Indian state, whose unity was being forged by a son of Gujarat, the '*Iron Man of India*', *Sardar* Vallabhbhai Patel.

Deprived of political power for many centuries, the clever and industrious people of Gujarat turned to trade. Their province was particularly suited for it, for they stood at the doorway to the Middle East, and their coast was dotted with a number of safe harbours where the big oceangoing Arab *dhows* could dock. The proverbial wealth of Gujarat was originally fostered by the trade in war horses, much of which was routed through the Gulf of Cambay (Khambat) and port cities such as Bhavnagar, Surat and Broach (Bharuch). Sultan Mahmud, with whom this history started, came all the way from Ghazni to raid the fabled wealth of the great temple of Somnath on the Gujarat coast. Thus, the wealth and power of the Gujarati people rested, as it still does, on the business, or *baniya* community. It was from this community that the greatest son of Gujarat was sprung. His name was *Mahatma* Gandhi.

To such a *baniya* family was Vikram Sarabhai born in August 1919. His father, Ambalal, was a man of no mean achievements. The son of a cotton mill owner, he travelled to England to learn the latest techniques in cloth manufature, which he applied to his Calico Mills and later the Jubilee Mills he inherited from his uncle. The business flourished, and it enabled Ambalal to expand into a new field every decade. In the 1920s it was textiles, in the 1930s it was edible oils and soaps, in the 1940s it was pharmaceuticals. The Sarabhai Chemicals, one of India's most respected makers of pharma products, was his creation. Vikram was the fourth child of Ambalal and his wife Saraladevi. Even more than Bhabha, he was born with a silver spoon in his mouth.

The young Sarabhai showed his talents early. He went to a private school called 'The Retreats' and was easily its star pupil. That fact that the school was owned by his family did not seem to give any privileges to the young prodigy. The atmosphere of the school was modern, and it instilled in him a lifelong curiosity in the way things work. After leaving this school, he enrolled at the Gujarat Arts and Science College in Ahmedabad, popularly known as the Gujarat College. This institution, which could well have figured in Chap. 3, had been founded as early as 1850 by Theodore Hope, an English civil servant (a product of the Haileybury school), who was among the first writers of elementary textbooks in the Gujarati language. It was and remains a premier institution of learning in Gujarat. However, after passing his intermediate examination from there Sarabhai did not continue. His father, who gave his brilliant son all the encouragement he needed, decided to send him to Cambridge for his graduation. The family could well afford the costs.

Enrolling at St. John's College, Cambridge, hard by the rooms where the great Newton had lived and a few minutes walk from Bhabha's Gonville and Caius College, Sarabhai passed the Tripos in Natural Sciences in 1940. Bhabha had taken the more prestigious Mathematical Tripos, but the Natural Sciences Tripos was no cakewalk either. In any case, it did not matter, for research ability, as is well known, is not always calibrated by one's ability to do well in examinations.

The World War was on, but in its first phase. Nothing much was really happening, but the storm was soon expected to break—and it did in May 1940. It was time for the young man to go home to India.

6.10   Bangalore Years

Sir C.V. Raman, with his usual flair for spotting an extraordinary talent, immediately brought the fresh graduate straight to Bangalore to join the cosmic-ray unit at the IISc. The fact that he knew the elder Sarabhai well may have helped.

The supremely confident youth worked in the same place as a luminary like Bhabha with mutual respect, but without getting drawn into his orbit. In this, he had the full support of their mentor, Raman. It may well have been Raman who encouraged Sarabhai to work on cosmic rays. investigating the latitude effect rather than the composition of cosmic ray showers which Bhabha was investigating. Sarabhai jumped into the subject with great enthusiasm—but he worked alone. Six single-author papers on this topic came from Sarabhai during his Bangalore period during which he studied the diurnal and annual variation of cosmic ray intensities and ways to separate the primary from the secondary cosmic rays. His main instrument, unlike Bhabha's devices, was the simple GM counter. It was, as has been described earlier, the cosmic ray physicist's main tool before photographic emulsions became the rage.

Sarabhai's very first paper was called 'Time distribution of cosmic rays' and it was published in the Proceedings of the Indian Academy of Sciences in 1942. It sought to find any systematics in the flux of cosmic rays as a function of time, and reported that there was nothing that could not be ascribed to random noise. As an aside, the paper acknowledges help from Kenneth Aston, whom we have met before in a more negative role.

In September 1944, the young Turk dared to provide an explanation for a phenomenon reported by his seniors Professors Aiya and Bhabha. They had reported a kink in the energy spectrum of cosmic rays. The supremely confident youth wrote a letter, which was published in Nature explaining that this was merely the point at which the flux of secondary cosmic rays takes over from the primary cosmic rays.

Earlier in the same year, Sarabhai had published two papers in the Physical Review, both entitled "The Method of Shower Anticoincidences for Measuring the Meson Component of Cosmic Radiation". In these he turned to the detection of muons in the cosmic ray showers. Instead of using absorbers to separate out the 'hard' muons from the 'soft' protons, as Bhabha and others had done, in the first paper he described an ingenious arrangement of GM counters with results recorded in anti-coincidence, which would do the same job. It was one of the early uses of the method which had been invented by Walther Bothe and would win him the Nobel Prize. Sarabhai's second paper describes his results with this apparatus, obtained by going up to heights as much at 13,900 ft in the mountains of Kashmir. More details of these Kashmir results had actually been published in a paper in the Proceedings of the Indian Academy of Sciences as early as January 1944. From these we know that the young scientist had taken his measurement near the lakes at Srinagar, Gangabal and Alpathar, the last of which took him to the neighbouring hill station of Gulmarg.

The last of the Bangalore papers was also published in the Proceedings of the Indian Academy of Sciences, and it was entitled "The semidiurnal variation of meson intensity". In this, Sarabhai returned to his first theme. He now had much more

sensitive apparatus to play with, and this time he found a faint but distinct oscillation
in the cosmic ray flux during one day, with a maximum during the afternoon and a
minimum at night. His interpretation that these are due to variations in atmospheric
pressure, however, proved untenable. These are, in fact, like the latitude effect, due
to the Earth's magnetic field, which undergoes a corresponding day-night oscillation
as the magnetic axis precesses around the geographical north–south axis. However,
this was not understood till the 1950s.

Though Sarabhai worked separately from Bhabha, the two became fast friends
across the decade of age difference. They had much in common—affluence, youth,
good looks, brains, memories of Cambridge and an appreciation of cultured femi-
nine company. It was thus in Bangalore that Sarabhai met his future wife. She was
Mrinalini Swaminadhan, younger sister of the Lakshmi Swaminadhan who would
soon become a Captain in the Indian National Army and fight for India's freedom
in Burma and Malaysia. Mrinalini was an accomplished Indian classical dancer, and
would become a famous one. She and others were to perform many times at Bhabha's
TIFR in the years to come, bringing to international conference audiences a flavour
of ancient Indian culture. A cultural programme showcasing Indian music and dance
thus became a tradition for Indian conferences for many decades, within and outside
TIFR.

In 1946, when the War was over and Bhabha and his team had moved to Bombay,
Sarabhai decided that it was time for him to go abroad again. Back to Cambridge
he went, and soon earned his doctorate (1947) with a thesis entitled *Cosmic ray
investigations in tropical latitudes*, which was based on his Bangalore measurements.
It also contained some results on nuclear fission, specifically the fission of Uranium-
238 nuclei by high-energy (6.2 MeV) gamma rays arising from the decay of Fluorine-
19.

6.11 The PRL

Armed with his new degree, Sarabhai came back to newly-independent India, all
of 28 years old. He did not want to go back to Bangalore, but wanted to found a
new Institute, just as Bhabha had done. When he mentioned this to Raman, the older
man asked him point-blank why he wanted his own Institute, when the I.I.Sc. itself,
and the new Institutes founded by Bhabha, M.N. Saha and K.S. Krishnan would all
be happy to have him and give him the facilities to work, as indeed he had worked
during the War years. Sarabhai's reply was characteristic. '*No plant grows under a
tree*,' he told the Nobel Laureate.

And that was that. Already, at Ahmedabad, in the premises of 'The Retreat', his
own *alma mater*, he had set up a cosmic ray laboratory, complete with a skilled
mechanic, an electronics expert and even a glass blower, under the aegis of the
Karmakshetra Educational Foundation set up by his own parents. This was to form
the nucleus for the new institution he dreamt of. He managed to enlist the help of

Kasturbhai Lalbhai, a rich industrialist who was one of the founders of the Ahmedabad Education Society. By an agreement between the two education platforms, it was decided in November 1947 to set up a Physical Research Laboratory (PRL) at Ahmedabad.

Sarabhai did not have the eloquence of Bhabha, and thus the objectives of the new Institute were stated baldly: *"The aim of the Physical Research Laboratory is to serve as a study and research centre for physics in western India and help to raise the standard of post graduate education in experimental and theoretical physics"*. The founder was always a strictly practical man. Not for him were the elegant lawns, the state-of-the-art buildings, the gorgeous paintings, and the sumptuous cafeteria which the stately Bhabha lavished on his brainchild. The PRL has always been strictly functional, even frugal, in its structure and decor. The equipment was moved from 'The Retreat' to the M.G. College[5] at Ahmedabad, and the PRL started almost immediately. Money started pouring in from the rich plutocrats who dotted Ahmedabad (and still do). Many of them were friends and relatives of the Sarabhai family. Gradually, funds also started coming in from the Council of Scientific and Industrial Research (CSIR) and Bhabha's Department of Atomic Energy (DAE), for Sarabhai was very much in touch with his old friends.

Though he was the founder of his new Institute, Sarabhai, at 28, did not want to become its Director. The post was filled by K.R. Ramanathan (1893–1984), a meteorologist who had once done a D.Sc. with Raman, but was then about to retire as Director of the Indian Meteorological Department. Sarabhai remained in PRL as a Professor, supervising no less than 19 students for Ph.D.'s in cosmic ray physics and related topics, including U.R. Rao (1932–2017) and K. Kasturirangan (b. 1940) who would successively preside over India's conquest of space from the 1980s till the early 2000s. Only in 1965, when Ramanathan retired at the age of 70, would Sarabhai become Director of PRL, a position which he retained till his untimely death in 1971. Unlike the formal Bhabha, however, Sarabhai always retained the common touch, and even when Director, was often seen in his lab in casual clothes, working with his students as if he were one of them.

At the PRL, Sarabhai went straight back to where he had left off in Bangalore, starting a major programme on measuring time variation of the cosmic ray flux. There is no room in this work for describing the more than 100 papers he brought out on cosmic rays and their physics. Suffice it to say that he based much of his research on the fact that measuring changes in the cosmic ray flux provides a means of mapping the Earth's magnetic field at high altitudes where one cannot easily ascend with a compass needle. And so Sarabhai's interests expanded from cosmic rays to planetary magnetic fields, from planetary magnetic fields to solar physics, from solar physics to satellite-based observation and eventually to the launching of satellites—at which point he determined that India should have its own space programme.

---

[5] For the curious, this is one M.G. in India which does not stand for Mahatma Gandhi. It stands for Mafatlal Gagalbhai, the founder of the Mafatlal group, one of India's leading textile makers, who was also one of the major benefactors of the Ahmedabad Education Society.

6.12   The Space Programme

The Space Age may be said to have commenced on October 4, 1957, with the launch of the first artificial satellite—Sputnik—by the Soviet Union. It also happened to be the International Geophysical Year. With the conquest of Space so much in the public eye, Sarabhai went ahead and proposed to the Government that India should have her own space programme. He was greatly encouraged in his idea of setting up a Space programme by Homi Bhabha, himself busily setting up an Atomic Energy programme. After knocking around in Government offices for five years, the proposal found favour, and thus the Indian National Committee for Space Research (INCOSPAR) was set up in TIFR in 1962 under the chairmanship of Sarabhai. In 1969, this became the separate Indian Space Research Organisation (ISRO), with Sarabhai continuing as Chairman. However, in the early years, the hub of activity continued to be the PRL, where Sarabhai worked immensely hard, continuing his research activity amid the setting up of a whole new programme. When one adds to this that Sarabhai was also the founder of a host of other institutions, from the iconic Indian Institute of Management at Ahmedabad to the Electronics Corporation of India at Hyderabad and the Uranium Corporation of India at Jadugoda in Bihar, one can only be astonished at the energy and versatility of the man.[6]

The story of the India's Space programme—one of the biggest success stories of modern Indian science—would take us too far afield in the present work. However, as we have already noted, the founder was an intensely practical man. In his own words *"We do not have the fantasy of competing with the economically advanced nations in the explorations of the moon or the planets or manned space flight. But we are convinced that if we are to play a meaningful role nationally, and in the community of nations, we must be second to none in the application of advanced technologies to the real problems of man and society, which we find in our country"*. It was not the glamour of adventuring into Space which attracted his interest, but the host of practical benefits that could ensure, including remote sensing, weather forecasting and seismic surveying.

In 1966, after Homi Bhabha's tragic death in an aeroplane accident, Sarabhai took over his role as the Chairman of the Atomic Energy Commission. He was now Secretary to two important Ministries, Director of the PRL and the family-run Sarabhai Chemicals, and Chairman of innumerable Government committees as well. The burden was too much for one human being, however talented, to bear. It all came to an abrupt end when the 52-year old scientist, then at the height of his career, went to sleep one night and never woke again in this world.

Sarabhai's legacy was carried on by his students and associates. As an epilogue, it must be said that though the prolific cosmic ray work of Sarabhai will not be quoted forever, as will be the Cambridge-period work of Bhabha, the *Chandrayaan* and *Mangalyaan* projects which make India so proud today would never have taken place

---

[6] Indeed, one can note with wonder that Sarabhai was one individual and not a group of scientists working under a pseudonym, as the late Stephen Hawking is famously said to have remarked about the Russian Yakov Zeldovich.

if the youthful Vikram Sarabhai had not once looked up at the star-studded Bangalore sky and wondered where the mysterious cosmic rays were coming from.

Chapter 7
The Return of Bose

7.1 The Science Congress

The Indian Science Congress has been initiated in 1914 by two British chemists, J.L. Simonsen and P.S. Macmahon. Simonsen was at the Presidency College, Madras, and Macmahon was at the University of Lucknow. The first session took place at the Asiatic Society in Calcutta, with Sir Asutosh Mookerjee as President. Since then, it took place on an annual basis. In 1938, the Silver Jubilee of the Science Congress was planned on a grand scale, again in Calcutta. Though the earlier sessions had been largely attended by scientists working in India, this time several eminent scientists from abroad were invited, and the President was supposed to be none other than the celebrated Lord Rutherford. In fact, however, Rutherford died suddenly that year, shortly before the Science Congress was held, and so his place was taken by another famous Briton, namely Sir James Jeans. At the Congress, Rutherford's address, which he had sent on in advance, was read out by M.N. Saha.

Among the other invitees to the Congress were Walther Bothe (1891–1957), a student of Köhlhorster and later a Nobel Laureate, and Sir Geoffrey Ingram Taylor (1886–1975), doyen of the Cambridge cosmic ray group. Of course, Indian luminaries like Sir C.V. Raman, Meghnad Saha, S.N. Bose and D.M. Bose also attended. In the cosmic ray session, there was a lively discussion on detection of cosmic ray particles, in which Bothe and Taylor took a leading part. One of their suggestions was that cloud chambers could eventually be replaced by stacks of photographic plates as cosmic ray detectors. These would be cheaper by far, and much more robust and would work even in low pressure conditions, such as on balloons and on high mountains. The technique had already been developed in Austria by Marietta Blau (1894–1970) and her student Hertha Wambacher (1937).

Photographic plates had been invented by George Eastman in 1879, and were initially manufactured by his creation, the Eastman-Kodak Company. In these a glass plate was coated with a thin film of 'photographic emulsion'—a colloidal solution of silver bromide or silver iodide in gelatin—and exposed to light focussed by a

© The Author(s), under exclusive license to Springer Nature Switzerland AG 2021
S. Raychaudhuri, *The Roots and Development of Particle Physics in India*,
SpringerBriefs in History of Science and Technology,
https://doi.org/10.1007/978-3-030-80306-3_7

lens. The impact of light causes the unstable silver halides to decompose to metallic silver with a variable density corresponding to the intensity of the light falling on different parts of the plates. This metallic silver gets oxidised to black oxide of silver and hence a 'negative' image is formed. The plate is then washed with a solution of sodium thiosulphate (called photographers 'hypo') which removes the excess silver halide. In normal photography, this negative image is then used as the source of a new photograph, which, being a double 'negative', will be a 'positive'. It was thus that all the wonderful black and white photographs of the early twentieth century were created.

As early as 1910, Cambridge-trained Japanese scientist Suekichi Kinoshita, who initiated radioactivity studies in Japan, and was then was working at the Tokyo Imperial University, had shown that alpha particles leave fine tracks on photographic plates, correctly interpreting this as due to the disintegration of silver halide molecules when impacted by highly energetic charged particles. When developed and enlarged, these tracks could be clearly seen and used to investigate the properties of alpha particles, which was Kinoshita's chief interest. However, though Kinoshita published some pretty pictures taken by his method in 1915, it didn't really catch on, because a year later came the invention of the cloud chamber, which yielded much clearer pictures than the low-density photographic plates then in use.

In Vienna, Marietta Blau had initially worked with a manufacturer of X-ray tubes and had spent some time as a radiologist, teaching medical students how to take X-ray pictures. Thus, she had a thorough grounding in photography and photographic emulsions, more so than most physicists would have had. At the time, in the 1920s, artificial radioactivity had been discovered, and measuring and tracking radioactive emanations was the hottest research topic in nuclear physics. At the Radium Institute of Vienna in 1923, two colleagues, Gerhard Kirsch and Hans Petterson, were investigating such effects using scintillation screens coated with zinc sulphide—the same device used by Geiger and Marsden in the famous gold-foil experiment. It was apparently Pettersson who asked Blau, with her photographic experience, to try and verify their results using photographic plates as detectors. Blau succeeded and in 1925, published her first paper "*On the photographic effects of natural H-rays*", i.e. protons, in which she reported some blackening of the plates due to the impact of these particles. It is interesting that as motivation for using photograpic plates, the very first paper cited by Blau is D.M. Bose's 1916 paper in which he proved the ionising effect of protons in a cloud chamber. There is no reference to Kinoshita, whose work had apparently been forgotten. Instead, there is a reference to a French-language paper by H. Mühlenstein, where apparently some evidence of particle tracks in photographic plates had been reported. It must be noted that Kinoshita had already been able to print pictures in 1915, but Blau's early work had not reached that level of sophistication.

In 1927, James Chadwick, future discoverer of the neutron, but then only known as Rutherford's star pupil, visited Vienna, and showed that the work of Kirsch and Pettersson was deeply flawed and that their scintillation method resulted in a huge over-counting of the actual number. This ended the scintillation studies at Vienna. Pettersson, in fact, quit nuclear physics and moved into geophysics. But Marietta

Blau persisted with her photographic studies. Gradually she found herself able to take pictures. With her student Hertha Wambacher, she began to expose the plates to cosmic rays, and, in 1937, they were rewarded by being able to record a characteristic 'star' which results from a direct hit on a nucleus in the emulsion and its disintegration. Of course these had already been seen aplenty in cloud chambers, but this was a first in photographic emulsions.

We have seen how the news of Blau's work had reached Cambridge (where Rutherford may well have remembered the early work of his one-time protege Kinoshita), and was taken up by Taylor and Goldhaber, with some success. Meanwhile, in 1938, Austria had been taken over by the Nazis, and Marietta Blau, being Jewish, had been summarily dismissed. The harm it did to her career was never compensated. The rest of her story is a pitiful mixture of racial and gender discrimination, which is only now being unearthed. In fact, she was never able to get a regular position again, moving from one temporary assignment to another. Seldom since the Middle Ages has a first-class scientist been treated so badly. The subject of all this died of cancer in 1970—she had handled too much radioactive material without protection. Her tragic end was similar to that of her great idol, Marie Curie. A similar fate had overtaken her erstwhile student Wambacher in 1950.

The discussions at the Science Congress fell on fertile ground. They struck D.M. Bose, whose cosmic ray interests had so far revolved around the cloud chamber, so forcefully, that when the Congress was over, and the distinguished guests had left, he decided to put his student Biva Chowdhury to carrying out cosmic ray studies with photographic plates.

But they would not be looking for radioactivity. Instead Bose had a more ambitious plan. After twenty years he had come back to cosmic ray physics. He was determined to track down Yukawa's elusive 'mesotron'.

7.2 Tracking Mesotrons

Biva Chowdhury was born in 1913, the daughter of an affluent doctor belonging to an old *zamindar* family from the district of Hooghly, to the west of Calcutta. The family belonged to the Brahmo Samaj, the reform movement started by Ram Mohun Roy, and were open and liberal in their views. Without this support, the young Biva would never have been able to take up science—then almost purely a male preserve—as a profession. She was educated at the Bethune College in Calcutta, the best education that a woman could get at the time. She then got her M.Sc. from the University of Calcutta, where she must have been taught by, among others, D.M. Bose, C.V. Raman and P.C. Mahalanobis. In 1938, when the great *Acharya* J.C. Bose died, D.M. Bose became his successor as Director of the Bose Institute, and Biva Chowdhury joined him as his student. She was a strong-willed woman, determined to follow her instincts and take up the career she wished. The only thing she was never able to make up her mind on seems to have been how to spell her name. Sometimes she wrote her first name as Biva, and sometime as Bibha (which is more phonetic), and once even as

Biwa. Her surname too, ranged from Chowdhry to Choudhury to Chowdhury. Only one combination has been used in this work.

The War was about to break out, and the Calcutta duo had no access to high-flying planes. Nor did they know anything about flying balloons, as Bhabha did. But being in India had one advantage—that of providing the highest mountains in the world. And so, they travelled to the Darjeeling Himalayas, about 600 km north of Calcutta, to expose their plates to high-altitude cosmic ray showers. They chose three 'stations'. The first was Tiger Hill, standing at 8,500 ft, a popular tourist destination, where, on a clear day, Mt. Everest can be sighted, even as Andrew Waugh had done once upon a time. More remote was the 12,000 ft peak of Sandakphu, in the Darjeeling Himalayas, about 60 km from Darjeeling, and even more remote was Phari Jong at 14,500 ft, some 200 km away on the edge of the Tibetan Plateau. We shall refer to these collectively as the Sandkaphu experiments.

Travelling to these remote sites was not easy. There were no motorable roads and so the slow climb up and downhill had to be accomplished on mule back. Mules were also employed to carry the heavy stacks of glass plates, which were fragile and could easily be dashed to pieces if the carrying mule stumbled or chose to rear up. Bose was 54 when these experiments started and 60 when they ended. In the reckoning of the times, he was an elderly man. As for Biva Chowdhury, at that time such journeys to remote places must have been regarded as highly adventurous for a young unmarried woman from a *bhadralok* background. The enterprise and dedication of this *guru-shishya* pair must be given the highest possible respect.

In these Sandakphu experiments, Bose and Chowdhury exposed a stack of 'Ilford plates' to cosmic rays at some location where they would be sheltered from the ravages of the weather. Their manufacturer, Ilford Ltd. had been set up in 1879 and still exists (though it is now Japanese-owned). The 'Ilford plates' came in two varieties—the 'half tone' plates, which were coated on one side only—and the 'full tone' plates, which were coated on both sides. Obviously the full tone plates gave better resolution than the half tone ones, but they were also much more expensive. In any case, as soon as the World War broke out, all the full tone plates were requisitioned for war use, where they would prove invaluable in aerial reconnaissance of the kind used, famously, to sink the *Bismarck*. The experiments of Bose and Chowdhury were done, therefore, with half tone plates, and this made their finding much less accurate than they could otherwise have been.

Studying their exposed prints, Bose and Chowdhury found the usual 'disintegration stars', many tracks which were clearly due to protons and another set of tracks, which they immediately identified (1940) as the 'mesotron'—the exchange particle of Yukawa, which we today call the pion. A pion candidate had been discovered by Anderson and Neddermeyer in 1937, and Oppenheimer and Serber had hailed it as the Yukawa exchange particle. But a series of studies—one by Bhabha as described, but more notably by Bruno Rossi and the trio of Marcello Conversi, Ettore Pancini and Oreste Piccioni—had led to the conclusion that this particle interacted too weakly with nuclei to be the Yukawa particle. Bose and Chowdhury were quite sure, however, that they had discovered the Yukawa particle. A series of papers in *Nature* followed (1941–44), in which they discussed indications that the tracks were indeed those of the 'mesotron', and that its mass was indeed in the right ballpark. In 1945, Biva Chowdhury wrote her Ph.D. thesis on her discovery of the 'mesotron'—a truly landmark thesis, which sadly never got the recognition it deserved.

Did Bose and Chowdhury really discover the charged pion, which is Yukawa's mesotron? This issue has been discussed again and again, more so in recent times. A clear answer can be found by looking at the tables of data published by Biva Chowdhury in the Indian Journal of Physics in 1944, and later reproduced in her thesis. These are presented in the graph below. It is quite clear from the graph, where the muon mass and the pion mass are clearly marked, that what the duo had seen was clearly neither the proton, whose mass lies at 938.4 meV/c^2, nor the electron, whose mass lies at 0.5 meV/c^2. It was therefore, clearly a meson which they had spotted, and they were indeed the first to do so using photographic emulsion. In fact, the data sets on the left seem closer to the real masses than the later data. Bose and Chowdhury, in a 1942 paper, explain that the rise towards right was due to a correction made by them in the use of a formula, where it was Homi Bhabha who had pointed out to them that one must use, not the average value of the atomic number of the scattering nuclei in the emulsion, but the root mean square value. That this took them away from the genuine results indicates (with hindsight) the presence of some compensating error which fortuitously made the first set more accurate. In any case, if one takes an average, the data in the graph yield a 'mesotron' mass of 189 $\pm$ 200 MeV/c^2, which is a result of dubious accuracy—as indeed any experienced eye can tell just by looking at the graph.[1]

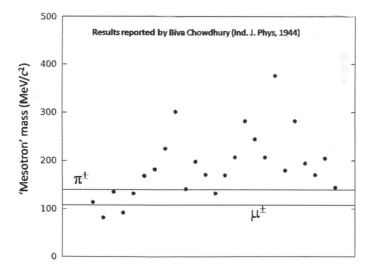

---

[1] An erroneus claim, based on averaging just the central part of the data set which lies close to the pion mass, that Bose and Chowdhuri's data did 'see' the pion, has been propagated in the media. Neither Bose, nor Chowdhury, whose scientific integrity was of the highest order, would have condoned such cherry-picking of data.

It is not as if the wide spread in the data did not bother the Calcutta scientists. In a 1944 paper, published in the *Physical Review*, they tried to understand this in terms of relativistic increase in mass, which would make the mass dependent on the kinetic energy of the 'mesotron'. As a justification, they tabulated results from other groups around the world, including famous names like Anderson and Nishina, which showed a corresponding spread of data. Today we know that cosmic ray pions are not really in the relativistic regime, and that the mass estimates mostly differed because of the primitive nature of the experiments themselves. This, at least, could not be blamed on Einstein!

The truth is that with their 'half-tone' plates, the experiments done by Bose and Chowdhury simply did not have the mass resolution to distinguish between muons and pions, which would be necessary to justify a discovery. It is not impossible that some of the tracks seen by them were, in fact, due to pions. However, as pions have a lifetime 80 times smaller than muons, this number could not have been more than a very few. In any case, the muon had already been discovered by Anderson and Neddermeyer, and it was not understood that the pion existed as a separate particle, with a rather similar mass. Not until this difference was resolved, could we say that the pion was discovered. This story will be related in the next section.

7.3   How the Pion Was Won

During the same period, the work of the Calcutta duo was being duplicated, among other places, at Bristol in the UK, by Cecil F. Powell (1903–69) and his group. Powell had been initiated into emulsion-based research in 1938 by none other than Walter Heitler, who knew Blau and Wambacher, and he (Powell) utilised the war years to work on this. However, as they were confined to Europe, and flying was impossible because of the Blitz, the highest point they could reach was the Jungfraujoch in Switzerland, less than the height of Sandakphu and considerably below Phari Jong. Thus the wartime results of Powell's group were pretty unremarkable, even though they also found tracks of mass intermediate between the electron and the proton, like everyone else.

In 1945, the War in Europe ended and most of the European and American scientists gradually drifted back from war work to their old research topics. Thanks to the vital contributions of scientists to the war effort (Radar, Bletchley Park, Tube Alloys Project), the British Government was very kindly disposed to science at the time. With Patrick Blackett as the chief scientific advisor to the Government, both Ilford and Kodak were instructed to produce special full tone plates with a high density of silver halide, and optical companies were instructed to produce better microscopes for plate scanning. The cosmic ray group led by Cecil Powell, at the H. H. Wills Laboratory of the University of Bristol, reaped the full benefits of this technology. Now, after the War, Powell was joined at Bristol by the dynamic young Brazilian César Lattes, who also brought into the collaboration his former teacher Guiseppe Occhialini, already famous for his work with Blackett on the 'coincidence technique', already quoted as the one used by Sarabhai in his Bangalore work. A fourth pillar of this group was Hugh Muirhead, then a doctoral student, who was later to write a particle physics textbook which was practically compulsory reading in the 1960s and 1970s. It was, however, Lattes who came up with the idea of adding some boron to the photographic emulsion. This had been done with success by Taylor and Goldhaber in the 1930s (as we have related), but no one had followed it up, not even those who initiated it. It is not known if Lattes had heard about this somewhere (for offbeat ideas have a habit of staying around in circulation for a long time), or came up with the idea on his own. But that it was adopted at Bristol at his insistence, there can be no doubt.

When the Bristol group exposed their new boron-impregnated high-density plates to cosmic rays, they were astounded to find a whole new world of tracks where none could be seen before. In the poetic words of Powell *"It was as if, suddenly, we had broken into a walled orchard, where protected trees had flourished and all kinds of exotic fruits had ripened in great profusion."* And among these exotics, they found the clue to the mesotron problem. There were actually *two* mesons, fairly close in mass, one of which decayed into the other. Today we write this reaction as $\pi^- \rightarrow \mu^- + \bar{\nu}_\mu$. Subsequently, the muon decays as $\mu^- \rightarrow e^- + \nu_\mu + \bar{\nu}_e$. We see the evidence of the charged particles π^-, μ^- and e^- in the famous 'golf-stick tracks' published by Powell and his group (1947) which established the existence of the pion as an independent entity. It was a rare moment of scientific epiphany.

The idea that there could actually be two different mesons had been independently suggested by some theoreticians as well. Shoichi Sakata, a student of Yukawa, who was naturally deeply interested in the 'mesotron' proposed by his mentor, had first proposed this in 1942, lecturing at the Annual Meeting of the RIKEN, the Japanese showpiece scientific centre of the times. This was some months after Pearl Harbour, however, and his audience was exclusively Japanese. Sakata followed this up by writing a paper on this theme with his student Takesi Inoue—in Japanese. The only person with which this struck a chord was Y. Tanikawa, and he wrote another paper with Sakata in 1947 on this topic. All these ideas remained confined to Japan, which, in 1947, was reeling under the impact of defeat and occupation. However, the same ideas occurred independently to Robert E. Marshak, who proposed them at the Shelter Island Conference of 1947, and then wrote them up in collaboration with Hans Bethe. By this time, however, the Bristol photographs were widely known and the Bethe-Marshak paper actually refers to them. Curiously, they refer to them as Lattes' results, with scant mention of Powell.

With the discovery of the pion, photographic emulsions became the standard tool for the discovery of new particles. Discovery after discovery followed, until, by the late 1960s, better results started coming in from accelerators, pushing cosmic rays to a niche in the broader field of particle physics. But in 1950, the Nobel Committee had already decided to award the Nobel Prize for *"...the development of the photographic method of studying nuclear processes and ... discoveries regarding mesons made with this method."* Their choice fell on Cecil Frank Powell, head of the Bristol group.

7.4   The Curious After-Life of Biva Chowdhury

There can be no question that Powell deserved the Nobel Prize. He had initiated a
major programme to find the pion, set up a group to do it, which then improved the
sensitivity of the method vastly, and been rewarded with success at the end of the day.
It is true that César Lattes had provided a crucial idea which helped in the discovery.
The actual track had been spotted by two women 'scanners' named Marietta Kurz and
Irene Roberts, who were non-scientists trained to spot unusual configurations in the
particle tracks. But the controlling brain had been Powell's. The Nobel Committee
recognised this, as its policy at the time was to award only the head of a laboratory
or project and not the assistants.[2]

One does not know the thinking behind the Nobel Committee's decision to make
the English physicist the sole awardee. The author is of the view that it would have
been fair to award the Prize jointly to three recipients—to Marietta Blau for discovering (actually re-discovering) the utility of photographic plates to study cosmic rays,
to D.M. Bose for proving that this could be used to track mesons, and to Powell for
resolving the meson tracks and proving the two meson hypothesis. The reason why
Blau and Bose were ignored, and the entire credit was given to Powell,[3] is but one
of the many conundrums created by the Nobel Committee in more than a century of
mostly uncontroversial awards.

Even if the Nobel Committee had awarded the Nobel Prize to Bose, however,
it would not have gone to Biva Chowdhury, who was merely his graduate student,
following his guidance and instructions. She did, however, get a Ph.D.—from the
University of Manchester. For supervising Biva Chowdhury's thesis work was really
D.M. Bose's last hurrah in cosmic ray work. He was 60 years old and had no appetite
for climbing mountains any more. Moreover, in the turmoil of the pre-Independence
period, which saw massive Hindu-Muslim riots in Calcutta, the safety of doing field
work was certainly in question. On Bose's advice then, Biva Chowdhury went off
to Powell's laboratory in Bristol, where, her mentor had assured her, she would
find much better quality equipment to work with. Powell, whose own laboratory
was saturated at the time, then found her a place with Blackett at the University
of Manchester, and there she finished her thesis and submitted it under Blackett's
official supervision.

The rest of Biva Chowdhury's career forms a long anti-climax to her pioneering
work in the War years. She came back to India to join Homi Bhabha's cosmic ray
group, but rather surprisingly, did not work with photographic emulsions at all.

---

[2] There are many examples of brilliant assistants being deprived for this reason. Examples are Morris
Travers, who assisted William Ramsey in the discovery of noble gases, K.S. Krishnan who actually
spotted the Raman effect, Jocelyn Bell-Burnell who spotted the first pulsar when working under
Anthony Hewish, and Giovanni Jona-Lasinio who co-proposed spontaneous symmetry-breaking
with Yoichiro Nambu.

[3] To be fair, Powell was always perfectly frank that his work was inspired by Marietta Blau's
discovery and encouraged by Bose and Chowdhury's early results. It is perhaps fitting that when
the telegram came to him that he had won the Nobel Prize, it was taken to his home by his young
Indian student—M.G.K. Menon.

Instead she built a cloud chamber and produced some elegant results. After a few years, she moved to the PRL, where Vikram Sarabhai gave her a warm welcome. There were no milestones in her career at Ahmedabad, except that she planned to set up a cosmic ray station at the nearby hill station of Mt. Abu. Sarabhai supported it strongly, but died before it could materialise. His successors were not so enthusiastic, as a result of which Biva Chowdhury took a premature retirement from the PRL and returned to her home town, Calcutta. Though she worked at the Bose Institute for some time, once again there are no milestones from this period. By the time she died, in 1991, aged 78, memories of her great work in the 1940s had mostly faded.

Over the next two decades, Biva Chowdhury became a forgotten figure. When the Indian Academy of Sciences brought out its landmark volume on Indian women scientists, named *Lilavati's Daughters* (2009) it did not feature her name. When the author of this work wanted a picture of the late scientist to exhibit in a talk at the centenary of the University of Calcutta's Physics Department (2016), there was none to be found on the Internet. It was the kindness of Professor Indrani Bose, of the Bose Institute, to dig up such a photograph, and that is the picture of Biva Chowdhury which is seen everywhere today.

However, the past few years have seen a remarkable revival of interest in Biva Chowdhury as an icon of the recent 'Women in Science' movement. Newspaper articles, social media posts, and a book called '*A Jewel Unearthed*' by Rajinder Singh and Suprokash C. Roy, all laud her achievements to the skies—in fact, literally so, for the star HD 86,081 in the constellation of Sextans has been named Bibha in her honour. She is now perhaps the best known woman scientist from India. In fact, by a supreme irony, it is now D.M. Bose who is the forgotten scientist and, as one sees every so often on the media, Biva Chowdhury who has become the 'woman scientist who was denied a Nobel Prize'. One wonders what the lady herself would have thought of all this. However, at the end of the day, these posthumous crumbs of recognition, though certainly welcome, form rather meagre compensation for what she was denied in her lifetime.

Chapter 8
The Growth and Spread of Particle Physics Research

During the 1950s, cosmic rays groups came up and flourished in the country and so did studies in the theory of elementary particles, one of which was being discovered every few months. With this spread completed, one could say that particle physics had taken root in the country and was no longer a delicate shoot sustained by the heroic efforts of a few individuals. This, then, forms the final part of our long story.

8.1 The TIFR Group

In 1945, when Bhabha moved to Bombay, many of his cosmic ray group moved with him. The first was his right-hand man, Ranjan Roy Daniel Nadar (1923–2005), generally known as Roy Daniel. A Tamilian from Nagercoil, the southernmost city in India, he was a product of the Loyola College and the Benaras Hindu University. He had worked with Bhabha at Bangalore on a balloon project and was, therefore, happy to join him at the newly-minted TIFR in 1947. His mentor promptly sent him off to Powell's lab, where he completed his Ph.D. with Donald H. Perkins (b. 1925), whose textbook on particle physics is still one of the most popular introductions to the subject. Daniel then came back to TIFR and remained there for the rest of his official career. Three bright young sparks who joined the group in those early days were B.V. Sreekantan (1925–2019), who was later to head the TIFR in the 1980s, Yash Pal (1926–2017), who later became Chairman of the University Grants Commission of India, and Devendra Lal (1928–2012), who would later become a prominent geophysicist and one of the builders of the Scripps Institution of Oceanography at the University of California at San Diego. Soon they were joined by S.Naranan and P.V. Ramanamurthy, who proved to be skilled experimental scientists.

Once settled in Bombay, Bhabha quickly roped in, from the Wilson College, the veteran H. J. Taylor, who became a Visiting Professor in the TIFR, and, in view of his age and experience, was made the head of the photographic emulsions group.

© The Author(s), under exclusive license to Springer Nature Switzerland AG 2021   119
S. Raychaudhuri, *The Roots and Development of Particle Physics in India*,
SpringerBriefs in History of Science and Technology,
https://doi.org/10.1007/978-3-030-80306-3_8

In 1949 came Biva Chowdhury, then a mature physicist of 36 years, returning from Manchester. In the 1950s, the group was enriched by the addition of M.G.K. 'Goku' Menon, fresh from his Ph.D. with Powell at Bristol, where he had discovered the anomalous $K \rightarrow 2\pi$ and $K \rightarrow 3\pi$ decay modes. These would eventually lead to the discovery of parity violation by Lee and Yang (1956). Menon led the balloon effort, with Roy Daniel, and would later go on to become Bhabha's successor as Director of TIFR. Sukumar Biswas, who had jointly discovered the Lambda hyperon with Victor D. Hopper in Melbourne (1950), came over from Calcutta, where he had been working after his return from Australia. The brilliant theoretician, E.C.G. Sudarshan, worked for three years (1952–55) in this group before moving to the United States permanently. Nowhere in India has such a collection of scientific talents existed except in the early days of the postgraduate science departments at Calcutta University.

The biggest catch of all, however, was Bernard Peters, originally Bernhard Pietrowski (1910–93), a Polish-born cosmic ray expert, whose parents had fled from Poland to Germany because of the Russian domination. Peters himself had to flee Germany because he was Jewish and had communist sympathies, escaping from certain death at the dreaded Dachau concentration camp, and he ended up in the USA, the Land of Freedom. Safe here, he continued his studies, eventually finishing his Ph.D. with J. Robert Oppenheimer at Berkeley in 1942. From 1942 to 45, he worked on the Manhattan project, but, like so many others, he was deeply upset by its horrific consequences. Moving away to Rochester, Peters involved himself deeply in cosmic ray studies, which was about as far from weapons research as a nuclear physicist could get. However, he soon found himself under the scanner of the then-ubiquitous Un-American Activities Committee, under the leadership of the sinister Senator Joe McCarthy. It was the time when the Rosenbergs were on trial for their lives, which they were soon to lose on the electric chair. Peters was a one-time communist, a nuclear physicist and a student of Oppenheimer to boot. And the maverick 'Oppie', when examined by the dreaded Committee, made all sorts of unguarded remarks about Peters' leftist views, driving the fanatical Senator into a frenzy of suspicion. It was beginning to look like Dachau again in the Land of Freedom. Fortunately, around this time, Peters met Homi Bhabha at a conference. Learning that Peters was headed for the notorious Blacklist, Bhabha promptly invited him come to India to join the TIFR and develop cosmic-ray observations there. In 1951, therefore, Peters moved to Bombay, out of the reach of Senator McCarthy and his sniffer dogs. Probably they were glad to see him go.

As a matter of fact, the cosmic ray work in TIFR in the 1950s is really the story of Peters and his group, for by then Bhabha was deeply involved in setting up the atomic energy programme. Peters and his team did solid bread-and-butter work in cosmic ray observations, but the great 1950s discoveries in particle physics—antiparticles galore, strange particles, neutrinos, parity violation, flavour mixing—were all made elsewhere. With hindsight, we can see that this was partly because Peters directed some of the effort away from elementary particles to cosmic-ray induced exotic isotopes, which was his primary interest, and some of these, like Beryllium-10,

Silicon-32 and Aluminium-26 were actually found.[1] After 1954, however, far higher fluxes could be obtained from nuclear reactors and the early accelerators. This led, for example, to the discovery of transuranic elements by Glenn T. Seaborg and his associates using the powerful Berkeley cyclotron. Cosmic ray research for exotic nuclei was, therefore, something of a research dead end.

The exotic nuclide programme at TIFR was gradually abandoned after Peters departed for Europe in 1959 to join the Niels Bohr Institute at Copenhagen. This led to a refocussing of the TIFR cosmic ray efforts on experiments in the Kolar gold mines, and it proved to be more of a success story. However, it would not be fair to blame it all on Peters. As Arthur C. Clarke famously said, discoveries in science are largely made by serendipity, and that quality has been largely missing in almost all of post-independence Indian science.

8.2 P.S. Gill and the Aligarh Group

While Bose, Bhabha and Sarabhai came from affluent families with close connections to the who's who of contemporary society, a less-known name, but no less a pioneer, who came from the opposite end of the social spectrum was Piara Singh Gill (1911–2002) usually abbreviated as P.S. Gill. Gill came from a poor farming background, where, as a child in the Hoshiarpur district of Punjab, he had to walk five kilometres and back every day to attend the Khalsa High School at Mahilpur. His elder brother was a committed nationalist, who was arrested and flung into prison by the British Government. This determined Gill to leave his own colonised country and seek his fortune in the New World after leaving school (1929). But he could only get a visa to Panama.

Arriving at Panama City, penniless but full of hope, he drove a taxi for 10 months and saved up enough money to take ship for San Francisco. Here he could walk in without immigration formalities, for at that time no walls existed in the Land of Opportunity. It was, however, the aftermath of the Great Depression and jobs were not easy to come by. Soon, however, the determined youth was able to get a Bachelor's degree from the University of Southern California, to finance which he did all kinds of odd jobs, including scrubbing floors, washing dishes and picking fruit. He then moved to the University of Chicago (1936), where he was able to do his Ph.D. under the supervision of none other than Sir Arthur H. Compton, the 1927 Nobel Laureate and discoverer of the famous Compton effect.

Working with Compton, Gill studied the well-known latitude effect at sea level, shuttling between America, Canada and Australia to do so. He was immensely popular in the student community, earning the nickname 'Pi' long before 'Pi' Patel appeared in world literature. In 1939, 'Pi' Gill presented a paper on cosmic ray distributions which was later hailed as the first indication of the spin of the pion, later established on a firm footing with much effort from proton-proton scattering.

---

[1] The naturally occuring isotopes are Beryllium-9, Silicon-28 and Aluminium-27.

This won him his Ph.D. in 1940 and he could have stayed on at Chicago as a research assistant. But Gill's mind was made up. He applied and secured a one-year travelling studentship and came back to India, where, at that time, there were practically no facilities for the kind of research he had been pursuing. The generosity of Compton allowed him to bring back with him the expensive equipment he had been using at Chicago.

By the time his stipend expired, Gill had found himself a job as a Lecturer at the Forman Christian College in Lahore, where he continued working on the 'directional distribution' of cosmic rays, which we would today call the zenith angle distribution. It was then that he would travel to Gulmarg, 8,700 ft up in the mountains of Kashmir, to make cosmic ray measurements. He was still using counters. Towards the end of the War, however, when the Japanese were in full retreat and there were whole squadrons of British war planes sitting idle at Indian air bases, Gill was able, with help from M.N. Saha, to persuade some of the bored pilots to take him up on high altitude flights where he could expose photographic plates to cosmic rays—for by then he had learned the new technology. We can recall that Bhabha did the same thing at Bangalore in the early years of the War.

It is characteristic of Indian science that during the war years, no fewer than four stalwarts were working on cosmic ray physics in India—Bose, Bhabha, Sarabhai and Gill—but they all worked independently, even though they were well aware of each others' work.

In 1946, Gill went on a lecture tour of the USA and Europe, where his work was very well received by senior scientists like Patrick Blackett, Pierre Auger, Viktor Hess (the discoverer of cosmic rays), and of course, his old colleagues at Chicago. The other important person to be impressed was Homi Bhabha, who wanted Gill to leave Lahore and join the TIFR as Professor of Experimental Physics. Gill, however, loved Lahore and would not come. And then, in 1947, came the Partition of India and the horrific riots that accompanied it. Gill's autobiography relates how he was still planning to stay on in Lahore, but was literally ordered by Prime Minister Nehru to quit Lahore immediately and join Bhabha's TIFR in Bombay. This order probably saved his life, for Gill's close associates who stayed back at Lahore were murdered in cold blood.

Though Gill did start some balloon-borne experiments at TIFR with skilled technician G. H. Vaze, he could not really fit into the group, being too ebullient and too independent-minded to accept Homi Bhabha's leadership. Within a year he had resigned and moved back to the USA. From there, in 1949, he was lured back to Aligarh Muslim University by the charismatic Vice Chancellor, Dr. Zakir Hussain (later to become the third President of India). At Aligarh, Gill built up his own cosmic ray group. It was then that he was able to set up, at his favourite hill station, the Gulmarg Research Observatory (1951). It was the first permanent station of its kind in India, and it was inaugurated by none other than Arthur Compton, who expressly travelled to India for this purpose.

At Aligarh, Gill set up all the apparatus for cosmic ray studies and started to train a large number of students, not just to make physics studies, but to build their own instruments. They included R.N. Mathur, S.P. Puri, T.H. Naqvi, M.K. Khera, I.S.

Mittra, A.P. Sharma, Y. Prakash, S.P. Hans and V.B. Bhanot. Like the cosmic ray group at PRL, and the Bombay group under Peters, the Aligarh group's interests veered away from elementary particles per se and became more involved in nuclear reactions, geomagnetism and shower profiling.

The lack of high quality scientific equipment in India has plagued experimental science in India since its inception. We have seen how it stunted the pioneering work of Bose and Chowdhury. It drove the mathematically-inclined Bhabha to design and fly his own balloons. It also affected P.S. Gill, who had carried his own equipment home from the USA. In 1963, therefore, he moved from Aligarh to become Director of the fairly new Central Scientific Instruments Organisation (CSIO) at Chandigarh, where he remained till his own retirement in 1971. Gill's dream was to see Indian experimental science become free from the shackles of imported equipment. The CSIO has, it is true, patented a fair number of instruments, but the sad truth is that most good Indian laboratories still import a good deal of their equipment.

After retirement, Piara Singh Gill tried his hand at entrepreneurship, setting up a company to manufacture magnetic heads for tape recorders. Eventually, however, he sold his business and went back to the USA to live with his daughter. There, he wrote his autobiography, where he raised many pertinent questions about the way Indian science has developed since independence. The ever-rebellious Sikh was particularly caustic about Homi Bhabha, with whom his personality clash continued till the latter's tragic death in 1966. In 2002, at the ripe old age of 91, Gill's own long innings came to a close. The questions raised by him remain.

8.3 Experiments at Kolar

Some of the oldest rocks in the Earth's crust are to be found in the Deccan region, where the volcanic action which killed off the dinosaurs also created vast convoluted lava fields, rich in minerals. Such a place is at Kolar, about 65 km north-east of Bangalore and it had been mined since the second century BCE, for gold lay in veins just under the surface and could be accessed by digging small pits. This gold contributed in no small way to the riches of the Ganga, the Chola and the Hoysala kings of yore.

Forgotten in the later mediaeval period, Kolar was rediscovered when a British surveyor reported in 1804 that local people were panning gold dust out of the mud from ancient mining pits which lay abandoned in the region. It was not till 1880, however, that John Taylor and Sons, of London, leased the land from the Maharaja of Mysore and opened up the Kolar Gold Fields (KGF) for full commercial exploitation. In the 1880s, there was a *desi* version of the Gold Rush at Kolar, with the population increasing five-fold in the next two decades. By 1905, there were a dozen different mines, all exploiting the so-called 'Golden Carpet', a vein of gold-bearing rock which extended for about four miles either way.

Of all these mines, the deepest was the Champion Reef, whose shafts eventually reached a vertical depth of 10,500 ft below sea level. It is still the deepest mine outside of South Africa, but now, alas! disused and derelict. In 1948, however, it was still active, and the significance of having a very deep mine so close to Bangalore had not escaped the eagle eye of Homi Bhabha. In 1948, he sent his student, B.V. Sreekantan, down the Champion Reef mine at Kolar with a sensitive GM counter to measure the cosmic rays flux at depths of 10,000 ft and more. The purpose was to see if, after passing through that much depth of solid rock, the cosmic rays consisted only of muons, or there were other components as well.

To assist him, Sreekantan soon had S. Naranan and P.V. Ramanamurthy, young students who had joined the cosmic ray group at TIFR after completing their M.Sc.'s. A good part of the electronics equipment was acquired at throwaway prices from the Chor Bazar, Bombay's notorious flea market, where a lot of military surplus equipment had ended up after the War. Sreekantan and his collaborators soon found that the flux of muons reduced dramatically as the depth increased and was likely to be completely shielded at depths of around 6,500 ft. However, at those depths, his Geiger counters began to tick again. Sreekantan reported this to Bhabha, who immediately guessed that this must be due to radioactivity in the rocks at this level. We may recall that in 1914, Watson and his team had not found any radioactivity in the surface rocks at Kolar. Sreekantan was then able to prove the correctness of Bhabha's hypothesis by putting lead shielding around his equipment, which stopped the alpha radiation from the radioactive rocks, but let the muons through. In fact, there are traces of radioactive Thorium-232 in the Kolar rock, especially at great depths.

Muon flux measurements won Sreekantan his Ph.D., but he and his team had to leave Kolar abruptly, as the British companies pulled out in 1954, claiming, literally, that 'there was no more gold in that mine.' But in fact, there was. The KGF mines got a new lease of life as the management was taken over from the private players, first by the Government of Mysore in 1956 (Kolar Gold Mines Undertaking), and later by the Goverment of India (Bharat Gold Mines Ltd.), as a public sector undertaking.

In 1955, the neutrino had been discovered, and its interaction cross-section with matter matched the inverse beta decay predictions made with the Fermi theory of weak interactions. In 1957, Moisey A. Markov, at Moscow, realised that the ideal place to look for neutrinos would be deep underground, since most of the noise from cosmic rays would be shielded out and only the neutrinos could penetrate that far. Markov did put a couple of his students to this job, but it never really took off till the Baksan observatory was initiated in the Caucasus mountains in 1967. Instead, Bruno Pontecorvo's suggestion of studying high fluxes of neutrinos at an accelerator gained favour, and ultimately led to the Nobel-winning experiments of Lederman and his team (1962) who proved the separate existence of the electron neutrino and the muon neutrino.

However, Bhabha, Menon, Sreekantan and their group were quick to take up Markov's suggestion, realising that the Kolar mines at depths greater than 6,500 ft were the perfect setting envisaged by the Russian Academician. Thus Menon, Sreekantan and Ramanamurthy returned to Kolar in 1960. They had now been joined by a new student, V.S. Narasimham. A collaboration was also set up with Saburo Miyake of Osaka University. First, some of the collaboration members—Narasimhan,

Ramanamurthy and Miyake—made careful measurements of the muon flux to determine the correct depth at which to place the neutrino detector. They found that at a depth of 8,800 ft, they did not detect a single muon even after running the experiment for two months over an area of roughly three square metres. When contrasted with the surface flux, which would be like a *million* muons per minute over the same area, it was clear that the shielding was really good. The actual experiment was, however, done at a depth of 7,500 ft in the Heathcote shaft of the Champion Reefs mine, mainly for logistic reasons.

The cross section of neutrinos with matter is so small that it would have required a detector weighing several kilotons to see neutrinos directly at the Kolar mines, and this was definitely outside the logistic and financial power of TIFR. However, the Kolar team came up with an ingenious solution. The miles of *rock* surrounding the mine shafts were as good as any detector, since neutrinos could easily interact with the nuclei in the rock through the inverse beta decay reaction $v_\mu + N \rightarrow \mu + N' + X$, producing muons. And the Kolar experiments were perfectly geared to detect muons. It would be easy to distinguish these secondary muons from cosmic ray muons, because the latter would be moving vertically, whereas these would come in laterally from the surrounding rocks. Thus, a heavy lead shield on top of the detector would ensure that no direct muons came in to create confusion. A tripartite collaboration was set up with the TIFR team, led by Menon and Sreekantan, the Osaka University team led by Miyake and the cosmic ray team at the University of Durham, led by Arnold Wolfendale.

There was formidable competition. Frederick Reines, then at the Case Institute of Technology, Cleveland, Ohio, and who, with Clyde Cowan, had discovered the neutrino in 1956, was known to be making plans for such an experiment in the EPR gold mines in South Africa in collaboration with the University of the Witwatersrand at Johannesburg. In fact, Reines had read Ramanamurthy's thesis on the absence of muon flux at Kolar and initially wanted to do his experiment there. He had written to Homi Bhabha about it and Bhabha had agreed on condition of an equal partnership, which Reines was unwilling to concede. He wanted TIFR to provide the logistics and leave the science to him. On learning this, Bhabha turned him down. And so Reines went to South Africa and set up his rival experiment at the East Rand gold mine near Johannesburg. It was, thus, a big challenge for the KGF group that they must get results before Reines did.

It was a highly sophisticated experiment for the times. Neon Flash Tubes (NFT), already in use at KGF in another experiment, and with which the Durham group had considerable experience, were used in large numbers to make tracking detectors. Osaka City University brought the magnet needed to build the magnetic spectrograph. The final setup consisted of five telescopes and two magnetic spectrographs. The first neutrino event was recorded in early 1965. And so, in 1965, the Kolar experiment was the first in the world to report the detection of atmospheric neutrinos. The paper was published in the *Physics Letters* of August 15, 1965, at time when India and Pakistan were engaged in a fierce war, and it was authored by C.V. Achar, M.G.K. Menon, V.S. Narasimham, P.V. Ramana Murthy and B.V. Sreekantan, representing TIFR, by K. Hinotani and S. Miyake, representing the University of Osaka, and

D.R. Creed, J.L. Osborne, J.B.M. Pattison and A.W. Wolfendale, representing the University of Durham. Reines was close on their heels. His paper, reporting the same discovery, was published in the *Physical Review Letters* two weeks later on 30th August, 1965. Bhabha could breathe easy again.

Another pioneering experiment done at Kolar was to use the detector as a neutrino telescope. Some neutrinos originate from extraterrestrial sources, like supernovae. Since neutrinos are not affected by the Earth's magnetic field, one can project the arrival directions of the neutrinos on the sky to look for the existence of such sources and search if a large number of events point to a particular region in the sky. The TIFR group tried to look for such sources, but did not find any worth reporting. The detector was actually too small to achieve this goal. Nevertheless, it was one of the earliest attempts to look for celestial neutrino sources and opened up the possibility of *neutrino astronomy*—now an important new field of research after the giant IceCube neutrino detector became operational in Antarctica from 2010.

In the 1970s and 1980s, the Kolar experiments were revamped to look for proton decay, an important prediction of Grand Unified Theories (GUTs). This study continued, with negative results, until the experiments were closed down in 1992. But all this belongs to a different era and will not be taken up in this work. It may be mentioned, however, that mining activities at Kolar came to a halt and the Bharat Gold Mines finally closed down the operation in 1992. At that point, the cost of bringing out the gold had become more than the profit to be made by selling it—and that is always the end of the road for any mining company. The early 1990s were a period of great economic turmoil for India, and the Kolar Gold Fields was just one of its casualties. With the withdrawal of human intervention, Nature quickly reclaimed the mines, which filled up with water when the pumps stopped working, and doubtless much of the labyrinthine network of underground shafts has collapsed by now. The derelict surface site remains today a jungle of crumbling buildings, rotting fixtures and decaying machinery—a sad spectacle of ruin and desolation.

The story of Kolar will not be complete without a mention of the mysterious 'Kolar events', which seem to have been due to the passage of weakly-interacting particles of masses around 3 GeV through the detectors. First reported in 1975, these have remained a mystery all through the past four decades. There have been recent speculations that these are due to the particles which make up the dark matter component of the Universe. If this is verified in any of the current dark matter detection experiments, then the first dark matter detection would have actually taken place at Kolar.

8.4   The Delhi University Group

It has been mentioned in a previous chapter that the real development of Delhi University took place after Sir Maurice Gwyer, a former Chief Justice of British India, took over as the Vice Chancellor of the University of Delhi in 1938. As the University had, till then, been functioning largely as an examining body, Sir Maurice, like Sir

Asutosh in Calcutta, decided to open postgraduate departments in the University. To start the Physics Department, he recruited none other than D.S. Kothari, whom we have encountered earlier at Allahabad. Kothari had gone to Cambridge in 1934, on a strong recommendation from Meghnad Saha. He had done his Ph.D. with R.H. Fowler, adding his name to the illustrious list which included Bhabha, Birkhoff, Chandrasekhar, Dirac, Eddington, Hartree, Mott and Lennard–Jones, and was now back to his position as 'demonstrator' in Allahabad University, at a monthly salary of Rs. 130 a month. From here, on another strong recommendation from his mentor Saha, he was appointed a Reader at the University of Delhi at the much more respectable salary of Rs. 350 a month. Around him, over the years, grew a galaxy of excellent teachers and researchers, including more students of Meghnad Saha, such as P.K. Kitchlew and R.C. Majumdar. In particular, the arrival of Majumdar sparked a flurry of theoretical particle and nuclear physics activity in Delhi University.

The Kothari-Majumdar collaboration dates to 1939, when they published a letter in *Nature*, pointing out that the 'meson'—or pion as we would say today—should decay into electrons, but that those had not been seen. In fact, pions decay prominently to muons, and the reason is due to the much heavier mass of the muons—a standard textbook exercise today. At that time Majumdar was in the Bose Institute, then under its new Director, D.M. Bose. By 1942, however, we find Majumdar at Delhi, publishing calculations of meson-electron interactions with a young student called Suraj N. Gupta (b. 1924), who later became famous for his formulation of gauge field quantisation. This collaboration continued even when Gupta went off to Cambridge and then to Dublin. However, it should be noted that the gauge field work was done by Gupta entirely on his own.

In effect, Kothari and Majumdar were the two pillars on which the Department of Physics and Astronomy at the Delhi University grew up. Apart from their research, they trained generations of physicists, among whom was Jogesh C. Pati (b. 1937), the first proposer (with Abdus Salam) of grand unification and proton decay. Kothari, like Bhabha and Sarabhai, was called to serve the Government of India, first as Chief Scientific Adviser to the Ministry of Defence and then as Chairman of the University Grants Commission, while his alter ego, Majumdar, ran the Physics Department. In both of these positions Kothari left a rich legacy. Interestingly, though busy with policy making, he would continue to teach his classes, and was loved and respected by generations of his students—or so his obituaries say.

A less suave, but no less charismatic person in Delhi was Samarendra Nath Biswas (1926–2005), better known as S.N. Biswas, or, in his later years, as *Dada*, meaning grandfather. Like his namesake, Sukumar Biswas, *Dada* went to Australia to do his Ph.D., which he completed from the University of Adelaide in 1958. His supervisor was Herbert S. Green, himself a student of Max Born, who specialised in statistical physics, but was also an expert on quantum field theories. Homi Bhabha brought this younger Biswas also to TIFR in 1959, but the effervescent Bengali youth chafed under Bhabha's dominance. In 1964, he moved to Delhi, and remained there till his retirement. Biswas died in 2005, sharp and characteristically caustic to the last.

The story of particle physics in DU after the days of Kothari and Majumdar is largely the story of Biswas and his colleague, compatriot and rival, Ashoke Nath

Mitra (b. 1929). A.N. Mitra was a student of R.C. Majumdar who went to Cornell University to do a second Ph.D. under Hans Bethe, though much of this work was done with Freeman Dyson. He later returned to India and, after a stint at Aligarh, eventually joined Delhi University in 1963. He is still an emeritus professor there, though now in his nineties.

Both Biswas and Mitra did sterling work in quantum field theories, current algebra and S-matrix theories in the days before gauge theories began to rule the roost. The sheer range and scope of the research they undertook and the problems they attacked are simply awesome. They were the heart and soul of particle physics in Delhi until the 1990s. Stories of their mutual rivalry are also legendary. But all that belongs to a different story.

8.5   Alladi Ramakrishnan and MATSCIENCE

The final larger-than-life character to appear on this stage is Alladi Ramakrishnan (1923–2008), who was a sort of Tamilian counterpart of Homi Bhabha and Vikram Sarabhai. He was the son of a legal luminary, Sir Alladi Krishnaswamy Iyer, who was the Advocate-General of the Madras Presidency for fifteen years and later a key member of the Constituent Assembly which created the Constitution of free India. The younger Alladi graduated from the Loyola College, Madras with Honours in Physics (1943). He then took a degree in Law and started out as a junior assisting his father, as many fledgling lawyers do. However, in his graduation year, he had heard an inspiring lecture by Homi Bhabha at the Presidency College, on Meson Theory, and this had kindled in him a burning interest which refused to be stifled under a pile of dry legal papers. After four years of the law office, he had had enough and decided to throw it up and become a scientist. Running into paternal opposition, like Homi Bhabha in a similar case, the young Alladi had to engage a highly persuasive lawyer to plead his case—his mother Lady Venkalakshmi. And so, in 1947, he quit legal practice forever and went off to work under his hero, Homi Bhabha, at the TIFR.

Joining TIFR when it was still functioning from the Kenilworth bungalow, Alladi was given a problem-with-a-difference by Homi Bhabha. It was to try to understand the development of a cosmic ray shower using stochastic methods, using a probability density function which Alladi would name '*product densities*'. This allowed him to model a shower development and quickly arrive at flux densities which Bhabha had calculated by a more *ad hoc* and laborious method. In fact, the heavy Monte Carlo simulations carried out by particle physicists today can trace their roots to this work of Bhabha and Ramakrishnan, though this is hardly recognised in the scientific literature.

Bhabha had a practice of sending off his brighter protégés to some of the top scientists in the world to finish off their doctoral studies—a sort of finishing school—so that they could come back to TIFR bringing new skills with them. In this way, Harish Chandra had been sent to Dirac, Yash Pal and Sreekantan had been sent to Bruno Rossi, Virendra Singh was to go to Geoffrey Chew and Alladi was to go to Maurice

S. Bartlett, the noted statistican, at Manchester. Not all returned to TIFR, though. The fledgling would often grow wings and take flight for different climes. Thus it was that Alladi Ramakrishnan, after getting his Ph.D. from Manchester, returned to the University of Madras, where he was appointed a Reader in the new postgraduate Department of Physics there. The brilliant crystallographer, G. N. Ramachandran, who had just become Head of the new Department at the age of 30, was quick to spot Alladi, aged 29, as a rising star. Unfortunately, the two stars could not get along and friction started rather early.

At the Rochester Conference of 1956, Alladi made the acquaintance over lunch of an elegant gentleman with whom he happily discussed physics and enthusiastically explained to him some of his own ideas about cosmic rays. This suave gentleman turned out to be none other than J. Robert Oppenheimer, father of the atomic bomb, who was then the Director of the Institute of Advanced Studies at Princeton. An invitation to spend a sabbatical year at the IAS followed, which Alladi was happy to take up (1957–58). On his return, all fired up to introduce into the rather sleepy University Department a series of lectures by stalwarts such as were the norm in Princeton, and also to revamp the antiquated syllabus, Alladi ran into the stone wall of Ramachandran and his cronies. Never one to mince his words, Alladi spoke his mind freely and consequently found himself transferred to the satellite Department which had been opened at Madurai, a temple town which had no scientific establishment to speak of.

Undaunted by this setback, Alladi would come back to his Madras home every weekend and there hold a 'seminar' with the students, who flocked around him. He would teach them quantum mechanics and relativity and quantum electrodynamics, none of which featured in the official syllabus. Lectures at the *Ekamra Nivas,* Alladi's family home (named after his grandfather Ekamra *Shastri*) were given, among others, by Donald Glaser, Murray Gell-Mann and Abdus Salam, all names to conjure with. Meanwhile Alladi had met and impressed politician C. Subramaniam, who would later become a Union Cabinet Minister. Subramaniam was sympathetic to his idea of creating an Indian version of the IAS at Madras, but nothing further materialised.

In 1960, came the turning point. The great Danish physicist Niels Bohr came to India on an extended visit and went around the country, meeting its scientists and watching them at work in his own quiet, but deeply observant way. Later, at a press conference, he was asked what he had liked about Indian science. The great man said that he had been impressed by the quality of the scientific discourse at Bhabha's grand institute at Bombay, but he was also impressed by Alladi's small, but intense group at his own house. This interview was reported in *The Hindu* next day, and it happened to catch the eye of the Prime Minister, Jawaharlal Nehru, who wanted to know more about this interesting development.

The publicity served Alladi well. He found himself suddenly transferred back to Madras. But the wheels of government run slow, and though he no longer had to make the journey to and from Madurai, his dream institute remained a dream. A year later, however, Maurice Shapiro, the American astronomer, visited *Ekamra Nivas*, and mentioned to C. Subramaniam, whom he met a few days later, that the way students clustered around Alladi was reminiscent of the way they clustered around

Oppenheimer in his teaching days. Subramaniam now managed to bring the Prime Minister himself to attend Alladi's 'seminar' on one of his trips to Madras. Of this visit, the Minister later wrote *"Jawaharlalji was greatly impressed by the enthusiasm shown by the students... and in particular to see four girls among the students. When the students told him that they needed an institution for the development of theoretical physics and mathematics, he asked me to examine the proposal and put up a note for his consideration."*

And so the Institute of Mathematical Sciences was inaugurated on January 3, 1962. Its official acronym is the IMSc, but it is widely known by its telex name MATSCIENCE. With this, India now had its fifth particle physics group, after Bombay, Ahmedabad, Aligarh and Delhi. The early days were unmistakably over.

Chapter 9
Observations and Reflections

The coming and spread of Western science in the Indian subcontinent, and its development in one specific area, viz. elementary particle physics, has been the theme of this work. Throughout, the purpose has been to give the reader a flavour of the story, which embraces a huge cast of characters and ideas, and a vast variety of social and cultural backgrounds to the different developments. Limitations of time and space make it impossible to cover all the scientific work done in a comprehensive manner, even in the limited area of particle physics. Nor is that the purpose of this work. Enough has been related, however, to draw some general inferences from the story, and perhaps throw open some of its aspects for discussion.

The first point which must be noted is that Western science came to India as a foreign idea, which the Indians wanted to learn and profit from. What may first have impressed Indian minds was European military technique. Acting in unison through drilling and hard discipline had been a key feature of European warfare from the Roman legions to the archers of Agincourt, and it was this feature, rather than better guns, which gave the Europeans the military edge which enabled the eventual conquest of the whole subcontinent. The British ascendancy in India also coincided with the Industrial Revolution. The Battle of Plassey was fought in 1757 and Hargreaves invented the Spinning Jenny in 1764. American freedom, which concentrated British aggression on India, came in 1776, the same year as James Watt's reciprocating steam engine was patented. Thus, as their grasp upon India became stronger, the colonials could bring more and more scientific and engineering marvels to overawe the 'natives'. Inevitably, this led to an overwhelming feeling of inferiority among educated Indians,[1] coupled with a determination to master the

---

[1] A humorous story, written in the early twentieth century in the vernacular, has the hero declaim *"What if I am enamoured of a foreign thing like the railways? The wheel will surely turn. One day we will drive shuttles to the stars and then we won't take the Britishers along, not for love or money."*

© The Author(s), under exclusive license to Springer Nature Switzerland AG 2021   131
S. Raychaudhuri, *The Roots and Development of Particle Physics in India*,
SpringerBriefs in History of Science and Technology,
https://doi.org/10.1007/978-3-030-80306-3_9

principles by which these marvels were being accomplished. We have seen this theme unfold in the first part of this book.

The role of religion and religious differences cannot be ignored in any story about India and her people. The British came into an India ruled by a thin crust of Islamic warlords, by whom the huge majority of Hindus was dominated, and, on some occasions, oppressed. To the Hindus, there was no difference between the Islamists and the Christians, and therefore, they were rather pleased to see their erstwhile masters bow down before the foreign conquerors. *Schadenfreude* is, after all, a feeling that knows no borders. Thus, in all the three centres of British power, it was the Hindus who quickly adapted to the ways of the new masters, and began to absorb their knowledge as fast as they could. This happened with the full connivance of the Europeans, for they entertained great hopes of converting the 'Hindoos' to their own religion. Experience gained through a dozen Crusades told them that it would not be much use trying to convert Muslims to Christianity, but when they looked at the ancient Hindu religion with its multiple layers of symbolism, starting from the crudest forms, it seemed to them a mass of shamanistic superstition, ripe for the triumph of the Cross. To their surprise, however, the joys of Heaven and the fires of Hell, preached to the 'natives' by the most eloquent padres, did not cut much ice with a people brought up to believe in the impermanence of life and the certainty of reincarnation. Something stronger was needed to attract the 'heathen', and Western science proved to be that trump card. The entire evangelical movement of the nineteenth century, which saw Western science take roots in India, was predicated on the idea that a scientific education would eventually bring the Hindus in droves to the folds of Christianity. But it didn't happen that way. The Hindus, belonging to a civilisation that has survived three millennia of invaders, managed to acquire the new knowledge without acquiring the new religion. Eventually the Muslims joined in. Finally it was the Europeans who had to depart, leaving the 'natives' to pursue science and to bicker among themselves, as they had always done.

A strong nationalist feeling dominated Indian attempts to acquire and practise Western science throughout the colonial period. Of course, scientists ultimately pursue science for its own sake, but there was a definite feeling among the early Indian scientists that they must prove to the supercilious foreigners that they could beat them at their own game. This feeling was put forward forcefully by Sister Nivedita, who wrote in 1910, *"The whole history of the world shows that the Indian intellect is second to none. This must be proved by the performance of a task beyond the power of others, the seizing of the first place in the intellectual advance of the world… Are the countrymen of Bhaskaracharya and Shankaracharya inferior to the countrymen of Newton and Darwin?"* It is clear from the names taken by her that by intellectual leadership, she meant scientific leadership.

One must not be amused or distracted by the effusive tone of the fiery Irish lady's remarks. In the nineteenth century, Europeans completely dominated the world, and sincerely believed (like Macaulay) that they represent a higher type of humanity than the people they subjected to the horrors of colonialism. This feeling showed up in their everyday interactions with Indians, raising up in every proud Indian breast an indignation which found expression in a fervent wish to surpass the Europeans in their own areas of excellence. It was particularly strong in Bengal, where Meghnad Saha and Sailen Ghosh actually numbered among the revolutionaries. But we must remember that there was a strong nationalist streak even in the Anglophile Bhabha. Indian science could not have progressed so rapidly without this powerful impetus.

Science was, and to some extent still is, a man's world. In the days when Western science was beginning to show its first pale shoots in India, it was considered completely out of bounds for a woman. This was mostly true even in Europe, where an Ada Lovelace, or a Sofya Kovalevskaya was considered a curiosity, and contemned as 'un-feminine' by society. Nevertheless, the emancipation of women made steady progress throughout the nineteenth century. At the beginning of the century, in England a woman was a mere chattel belonging to her husband and the ducking stool was still in use. By its end, women could already vote in New Zealand, and the suffragette movement had started in England. In India, suttee was rampant when the century started, but by its end, girls' schools were mushrooming all over India. In the middle of the century, Macaulay would mock at ancient Indian science, pooh-poohing "...*astronomy, which would move laughter* in girls *at an English boarding school...*", which makes it abundantly clear what his Lordship thought to be the lowest form of the superior European mind. But towards the end of the century, Kadambini Ganguly was a practising doctor in Calcutta, Chandramukhi Basu was Principal of the Bethune School and *Pandita* Ramabai's *Mukti Mission* had taken off at Poona. The progress of women towards equality in science and elsewhere has continued unabated till present times, often at an excruciatingly slow pace—and there is still a long way to go. However, we must remember that Indian society is not a monolith. There coexist in this ancient land the most primitive societies along with the most contemporary ones, the most conservative and obscurantist groups along with the most revolutionary and *avante garde* ones. This must always be kept in mind, if India is to be understood in any context.

Towards the end of the first half of this work, the focus shifts to Physics in particular, among the other subjects, though there was plenty of growth and many outstanding names in other branches of science as well. This merely reflects the lack of expertise of the author in other subjects, and no other opinion. A really notable omission is Srinivasa Ramanujan, perhaps the most original scientific mind ever, whose transcendental genius for mathematics would have seemed like a legend if it had not been so closely documented. However, even with the limited coverage in this work, it becomes obvious that by the second decade of the twentieth century, several Indian scientists were beginning to make their mark on the international scene. Outside Europe and America (with which Canada and Australia may also be included) it was only in Japan that there were comparable scientific developments.

The second half of this work looks at some of the highlights of particle physics research, as it developed in India from rather humble beginnings. Again, we must remember that although particle physics and cosmic ray research grew out of nuclear physics, there was always a parallel growth of nuclear physics itself during the period of interest here. A separate work would be required to do justice to the work of stalwarts like Meghnad Saha and his team who built the first Indian cyclotron, to K.S. Krishnan, who tried to understand what we would today call the strong interactions, to the seminal work of S.N. Ghoshal which proved the existence of the compound nucleus, and, last, but not least, the work of P.L. Kapur with Sir Rudolf Peierls on what we would today call narrow resonances in nuclear scattering. Scientists like Raman, Saha, Kothari and Majumdar were also amazingly versatile, and could contribute with ease to topics as diverse as stellar atmospherics to improvement of GM counters to the study of band structure in solids.

This story draws to a close as it reaches into the mid-sixties. By then, particle physics research was well established in India, with several centres and satellite centres stretching to the four corners of the country. The first euphoria of Independence had worn off, with the new students coming in having been born in a free country. The colonial masters had been replaced by new satraps, and some of the rebels of the 1930s were now the kings of the castle. Overweening bureaucracy had begun to penetrate into everything, and the brain drain was truly on its way, sapping the strength of the fledgling science community which the early researchers had built up with so much hope and effort. Not till the 1990s would the Indian science community begin a slow recovery from these years of Government parsimony and neglect of academics—and that recovery is still far from complete today.

The importance of Cambridge University and the Cavendish Laboratory as a training ground for Indian physicists cannot be overemphasised. Except for S.N. Bose and P.S. Gill, almost every worker of note mentioned in the story was trained at Cambridge, or trained by Cambridge-trained people. In fact, the quality and impact of the research decreased in inverse proportion to the distance from Cambridge. Thus, the most impactful work in cosmic ray research is definitely that of Bhabha while he was part of the Cambridge group, and perhaps the most ignored work was that of Bose and Chowdhury in remote Calcutta during the War. People in those days were aware of it, and we repeatedly find them writing that they wanted to found institutions which would replicate the atmosphere of Cambridge and the Cavendish lab. This is a common thread which runs through the letters and speeches of everyone, from Maurice Gwyer to Bhabha, to Sarabhai, to Alladi Ramakrishnan. However, there is a fallacy in all this. Cambridge *grew*, and that too over centuries. The pioneers in India were therefore in the position of someone who plants a cutting and waters it furiously, hoping that it will grow overnight into a mighty tree. This is not to detract from their efforts, which were truly Herculean, but just to set our expectations at the right level.

The other fact which repeatedly strikes one is that almost as soon as an Indian scientist became a figure of international repute, he quit science and became a manager. C.V. Raman, D.M. Bose, Meghnad Saha, Homi Bhabha, Vikram Sarabhai, D.S. Kothari, P.S. Gill and Alladi Ramakrishnan, who feature in this story, all succumbed to this impulse. The pattern has been replicated across subjects, and across the years. Of course, one can argue—and it has been so argued—that these brilliant minds had to face so many obstacles themselves that they decided to sacrifice their own scientific growth in order to ensure that the next generation of brilliant minds did not face the same obstacles. This is a fine and noble sentiment, but not necessarily the most practical one. For these brilliant minds were obviously doing fine, even with all the obstacles—else they would not have been the famous scientists they were. Their heroic sacrifice would have been worthwhile if there were other, equally brilliant minds among their successors, who could have reaped the advantage of their self-sacrifice. As it turned out, the next generation had no dearth of competent workers, but the spark of their *gurus* was missing. To quote the outspoken E.C.G. Sudarshan, "*... it is a sad commentary that the momentum of Bhabha's discovery* [Bhabha scattering] *did not inspire comparable work in India... There was no close group of brilliant young colleagues and pupils who made up a critical mass to sustain his spirit.*" To some extent, the pioneers themselves were responsible for this, as they often showed themselves intolerant of disputation. Thus, there was some truth in Vikram Sarabhai's retort that plants in the shadow of a great tree remain stunted.

The poor participation of women in this game, at least in India, is another striking feature of the story. Except Biva Chowdhury, no other woman even finds a mention. The story of Marietta Blau shows that gender parity was still a far cry, even in the West. In that sense, over the years, there has been an explosion of women in science in the world and in India, though, of course, the numbers are still small. Until women are free—and feel free—to pursue careers in science, the brains of one half of the population will remain largely untapped. One is reminded of a verse from the *Hitopadesha*, which reminds us that a chariot cannot run unless both its wheels can turn freely.

Perhaps the most important lesson to be learnt from this historical excursus is that in the first part of the twentieth century, Indian scientists were doing research at the absolute cutting edge of contemporary knowledge. They fearlessly competed with giants, threw up new ideas without fear of ridicule and were willing to do the most difficult experiments with the most primitive of equipment. The number of original scientific ideas which flowed out of Indian minds has not been replicated since. After independence, India founded many new institutions, with excellent teaching—the IITs come to mind—and imported supposedly healthy academic practices like peer review and tenure track positions to ensure that our burgeoning science community stays on track. And that is precisely what has happened—Indian scientists have stayed on the track, some going fast and some going slow—but other people have been *laying* the tracks. In the 1930s and 1940s, there were no tracks, and so Indians laid their own.

This brings us to the final point, which is the question asked most often to Indian scientists by laypersons. Why is it that after Independence, Indian science, now supported by the Government, and free of the curbs put on it by colonial adminis-trators, has failed to produce giants such as it did in the first half of the twentieth century? One can identify many minor reasons for this, including poor salaries and the brain drain, academic timidity, empire-building by important scientists, and so on. However, these are symptoms, rather than reasons, and much of these go on elsewhere as well. The author is of the view that the real reason lies in the changing nature of scientific research over the past half-century. Earlier, science had been prin-cipally driven by *individual* effort, and the story of early-twentieth century science is replete with instances of heroic efforts made by scientists working in isolation, or with an assistant or two. This was essentially a nineteenth-century model, when a scientist was always a teacher—a Professor—who used his spare time after lectures to work in his laboratory. The importance of cooperative and collaborative work was first realised because of the great success of the Manhattan Project in creating atomic weapons in record time. The corporatisation of science has followed, with research becoming a commodity for industrial-style production, rather than a pursuit of puzzles for their sheer intellectual challenge. Thus, major advances in science are now rarely made by individual scientists working in small laboratories (though a great deal of bread-and-butter research does happen that way), but by large collaborations where an assembly-line style of operations is followed. Nothing brings this home better that the fact that the Nobel Prize, a nineteenth century creation awarded for individual merit rather than collective achievement, is nowadays mostly awarded to octogenarians, whose great work was done in a more heroic age.

It is precisely in this feature that Indian science—and the Indian attitude to science—lags behind. Just as the European invaders were much better at getting drilled soldiers to fight in disciplined lines, the advanced nations have largely been able to marshal their scientific talent into groups and collaborations, which are more effective than individual efforts. In India, however, this has been slow to take off. People still look back wistfully at the past and mourn the lack of towering scientific figures. Senior scientists bicker among themselves and permit their personal likes and dislikes to submerge their scientific instincts. Every prominent scientist, to take Sarabhai's analogy further, wants to be a lone-standing tree and not part of a forest. Yet, it is not scientific heroes who are required today, but the need to pull together and work efficiently in groups. After all, a battalion of well-disciplined troops would have made short work of an Achilles or an Arjuna. It is to this end that Indian science must strive.

Bibliography

Further Reading-Written by the Author[1]

1. The early days of particle physics in India, published in Proceedings of 'CU Physics 100' (2016), a conference to celebrate the centenary of the Department of Physics at the University of Calcutta
2. The origins of western science in India, published in Proiti (2017), a special volume in commemoration of the 150th Anniversary of Sister Nibedita
3. With Mondal NK. : M.G.K. Menon—statesman of Indian science, published in Resonance 24(11):1189 (2019)
4. *Bharotey bigyan shikkha o gobeshona—kichu elomelo chinta* (in Bengali), published in the Autumn Annual by the Presidency College Alumni Association, vol XLIX, pp 106 (2020)
5. Cosmic ray research in India—a historical perspective, published in Physics News (2021), special volume in memory of Biva Chowdhury

Further Reading[2]

1. Al-Beruni, *Kitab-al-Hind* or The Book of India, translated by Edward Sachau (1910)
2. Rennel J (1900) The journals. Asiatic Society. Some of Rennel's maps may be viewed at https://apps.lib.umich.edu/online-exhibits/exhibits/show/india-maps/rennell
3. Keay J (2000) The Great Arc. Harper Collins
4. Mukhopadhyay U (2013) Radhanath Sikdar. Dream 2047 15(10):28
5. Chaudhuri S (ed.) Calcutta - The Living City. Oxford University Press
6. Chakrabarti P (2004) Western Science in Modern India. Permanent Black
7. Kocchar R (2011) Hindoo College Calcutta revisited. In: Proceedings of the Indian history congress
8. Masani Z (2012) Macaulay: Pioneer of India's Modernisation. Random House, India
9. O'Connor JJ, Robertson EF. Ramchundra, copiously linked MacTutor website available at https://mathshistory.st-andrews.ac.uk/Biographies/Ramchundra/
10. Sharma RN, Sharma RK (2004) History of Education in India. Atlantic Publishers
11. Kocchar R (2021) English Education in India 1715–1835. Routledge
12. Adriaan G (2011) Mahendralal Sarkar. Brev Publishing
13. Arnold D (2004) Science, technology & medicine in Colonial India. Cambridge University Press

© The Author(s), under exclusive license to Springer Nature Switzerland AG 2021   137
S. Raychaudhuri, *The Roots and Development of Particle Physics in India*,
SpringerBriefs in History of Science and Technology,
https://doi.org/10.1007/978-3-030-80306-3

14. Kumar A (1998) Medicine and the Raj 1835–1911. AltaMira Press
15. Bose DM (ed.) (1971) A concise history of science in India. INSA
16. Chakravorty A (2014) The chemical researches of Acharya P.C. Ray. Ind J Hist Sci 49(4):361
17. Ray PC (1903) Life and Experiences. Chuckervertty, Chatterjee & Co
18. Geddes P (2000) The life and work of Sir J. C. Bose (1903). Reprinted by Asian Educational Services
19. Sarkar TK et al (2005) History of Wireless. Wiley-Interscience
20. Sengupta DP et al (2009) Remembering Sir J.C. Bose. World Scientific
21. Mukherji P, Mukhopadhyay A (2018) History of the Calcutta School of physical sciences. Springer
22. Walter M (2013) Early cosmic ray research with balloons. Nucl Phys B (Proc Supp) 239:11
23. Dorman LI, Dorman IV (2014) Cosmic Ray History. Nova Science Publishers
24. Born M (1978) My Life. Routlegde
25. Singh V (2009) Bhabha's contributions to elementary particle physics and cosmic ray research. Published as e-Print: 0905.2264 [physics.hist-ph]
26. Deshmukh C (2010) Homi Jehangir Bhabha. National Book Trust
27. Chowdhury I, Dasgupta A (2010) A Masterful Spirit. Penguin Books
28. Shah A (2007) Vikram Sarabhai—a life. Penguin Books
29. Brown LM et al (ed.) Pions to quarks. Cambridge University Press
30. Singh R, Roy SC (2018) A Jewel Unearthed. Shaker Verlag
31. Gill PS (1992) Up Against Odds. Allied Publishers
32. Ramakrishnan A (2019) Alladi Diary. World Scientific
33. Aggarwal V (2018) Leading Science and Technology: India Next. Sage Publications

Printed in the United States
by Baker & Taylor Publisher Services